T0137275

Springer Theses

Recognizing Outstanding Ph.D. Research

Aims and Scope

The series "Springer Theses" brings together a selection of the very best Ph.D. theses from around the world and across the physical sciences. Nominated and endorsed by two recognized specialists, each published volume has been selected for its scientific excellence and the high impact of its contents for the pertinent field of research. For greater accessibility to non-specialists, the published versions include an extended introduction, as well as a foreword by the student's supervisor explaining the special relevance of the work for the field. As a whole, the series will provide a valuable resource both for newcomers to the research fields described, and for other scientists seeking detailed background information on special questions. Finally, it provides an accredited documentation of the valuable contributions made by today's younger generation of scientists.

Theses are accepted into the series by invited nomination only and must fulfill all of the following criteria

- They must be written in good English.
- The topic should fall within the confines of Chemistry, Physics, Earth Sciences, Engineering and related interdisciplinary fields such as Materials, Nanoscience, Chemical Engineering, Complex Systems and Biophysics.
- The work reported in the thesis must represent a significant scientific advance.
- If the thesis includes previously published material, permission to reproduce this must be gained from the respective copyright holder.
- They must have been examined and passed during the 12 months prior to nomination.
- Each thesis should include a foreword by the supervisor outlining the significance of its content.
- The theses should have a clearly defined structure including an introduction accessible to scientists not expert in that particular field.

More information about this series at http://www.springer.com/series/8790

Adrian Love

Hollow Core Optical Fibre Based Gas Discharge Laser Systems

Doctoral Thesis accepted by
the University of Bath, Bath, UK

 Springer

Author
Dr. Adrian Love
Department of Physics
University of Bath
Bath
UK

Supervisor
Prof. William J. Wadsworth
Department of Physics, Centre for Photonics
    and Photonic Materials
University of Bath
Bath
UK

ISSN 2190-5053   ISSN 2190-5061   (electronic)
Springer Theses
ISBN 978-3-030-06759-5   ISBN 978-3-319-93970-4   (eBook)
https://doi.org/10.1007/978-3-319-93970-4

This Springer imprint is published by the registered company Springer International Publishing AG
part of Springer Nature
The registered company address is: Gewerbestrasse 11, 6330 Cham, Switzerland

For Mum, Dad,
Cerri, Sam,
and Lorelei

Supervisor's Foreword

Gas lasers were amongst the earliest laser types to be demonstrated and commercialised, and the red helium–neon laser quickly became the default for low-power visible laser light. The helium–neon laser is one of a large family of neutral noble gas lasers which share many common features in their performance and the dynamics of operation. In particular, the electrical discharge that drives the laser action was found to give higher gain for small laser tube diameter. However, it was also recognised in the early 1970s that there is a limit to which gas lasers can benefit from really small diameter tubes, which is set by the fundamental physics of diffraction of the laser beam. If the bore is too small, then the diffracting laser beam will hit the inner sides of the gas tube and will usually be lost. This balance for tubes of small diameter, d, between increased gain (proportional to $\frac{1}{d}$) and the increased diffraction loss (proportional to $\frac{1}{d^3}$) led to lasers with typical inner diameter of 0.5 to 1 mm, and operation at 0.25 mm diameter demonstrated in short tubes.

Hollow optical fibres present a glass tube where light incident on the inner walls is reflected back and trapped in the central hole. In 1999, a hollow optical fibre for visible light using the principle of a photonic bandgap was demonstrated and immediately opened up the possibility that the long-held diffraction limit could be broken in gas lasers. These early fibres had hollow bore diameters of only a few micrometres, which represented more than an order of magnitude reduction in bore diameter. The potential increase in laser performance was, however, difficult to attain as sustained electrical discharges are problematic in such extremely small tubes, and it was not until 2008 that we started experiments towards discharges for gas lasers in small tubes. This initial research indicated that a bore diameter of about 100 μm would be required, and that operation of lasers in the mid-infrared (using xenon or helium–neon) would offer the best opportunity of success. Unfortunately, these parameters were not available in any of our fibres at that time. In 2011, a new class of large bore hollow optical fibres opened up the opportunity to simultaneously confine light and allow an electrical discharge. These optical fibres are remarkable in themselves, they are made from silica but offer high transmission at 3.5 μm wavelength where silica is very highly absorbing.

Adrian has made this breakthrough by careful consideration of the implementation of the discharge and by developing the measurement and analysis to demonstrate laser action in an optical cavity. Adrian implemented a mid-infrared laser in xenon in a 120 μm diameter flexible optical fibre core which, at 1 m length, is several hundred times the diffraction limited Rayleigh length.

Bath, UK   Prof. William J. Wadsworth
April 2018

Abstract

The humble electrically pumped gas laser has undergone little development in its 50-year lifespan due to the lack of an effective method to confine light within a hollow waveguide of any appreciable length in which an electrical discharge could be contained. New technologies in the field of anti-resonant guiding hollow core fibres present an opportunity to re-invent the gas laser. A recent breakthrough in the field demonstrated that DC pumped glow discharges of a helium and xenon gas mixture could not only be sustained in such a fibre, but also exhibited signs of gain on a number of mid-IR neutral xenon laser lines.

The research presented in this thesis is a continuation of that project. The system was redesigned to incorporate two mirrors so that a cavity could be constructed. The previously hinted at gain on the 3.51 µm xenon line was confirmed through a series of CW measurements of the cavity, as was a polarisation of the laser due to a polarisation-dependent output coupler.

Further observation of the discharges revealed that they were of a pulsed nature, and that the mid-IR laser light was present in the discharge afterglow. A response to the cavity mirrors was observed in this afterglow pulse on the 3.11 and 3.36 µm xenon lines in addition to the 3.51 µm line previously seen. Through fast detection, a modulation of the output power due to cavity mode beating effects was detected. The high gain and narrow bandwidth of the xenon laser lines resulted in a frequency pulling effect, and the mode separation in the 'hot' laser cavity was measured to be lower than in the 'cold' cavity.

It was observed through pressure optimisation experiments in helium–xenon that higher output powers could be achieved by using lower partial pressures of xenon. This was exploited with neon–xenon mixtures, where the lower ionisation potential of neon allowed a lower pressure of xenon. Discharges were also achieved in helium–neon and argon gas mixtures.

List of Publications

Publications

1. A.L. Love, S.A. Bateman, W. Belardi, C.E. Webb and W.J. Wadsworth, *"Double pass gain in helium-xenon discharges in hollow optical fibres at 3.5 μm"*, in *CLEO: 2015*, SF2F.4, Optical Society of America (2015).
2. A.L. Love, S.A. Bateman, W. Belardi, F. Yu, J.C. Knight, D.W. Coutts, C.E. Webb and W.J. Wadsworth, *"Gas lasers beyond the diffraction limit"*, *In Preparation* (2017).

Featured Presentations

1. A.L. Love, S.A. Bateman, W. Belardi, C.E. Webb and W.J. Wadsworth, *"Double pass gain in helium-xenon discharges in hollow optical fibres at 3.5 μm"*, in *CLEO/Europe: 2015*, CJ-5.1, Optical Society of America (2015).
2. W.J. Wadsworth, A.L. Love and J.C. Knight, *"Hollow-core fiber gas lasers"*, in *Workshop on specialty optical fibers and their applications*, WT1A.1, Optical Society of America (2015).
3. W.J. Wadsworth, A.L. Love F. Yu, M.R.A. Hassan, M. Xu and J.C. Knight, *"Gas filled hollow core mid-IR fibre lasers" (invited)*, in *CLEO/Europe: 2017*, CJ-8.1, Optical Society of America (2017).

Acknowledgements

At the time of writing this, it has been 3 years, 9 months and 24 days since I started this doctorate, and in that time I've built up a large debt of things to be thankful for. So I better get on with it.

First and foremost, I would like to thank my supervisor William for trusting me with this project, for all the guidance given, for the outreach opportunities (especially the RSSSE) but most of all for giving a lowly third-year undergraduate from a different university a summer placement opportunity. I will be forever grateful.

Next, I'd like to send my thanks to any other members of the CPPM past and present who may have helped me over my time in the group. Particularly to Pete for doing what every second supervisor should; continually buying me drinks, and to Tim for being our sounding board in what must have been some of the strangest meetings he's had. I'd also like to thank Sam for finishing just as the project got interesting and Steve for his help in my brief stint as a fibre fabricator.

There are also a number of support staff outside the CPPM who need thanking. So a big thanks to Paul for his workshop skills on any project, big or small. That ceramic cell can't have been easy. And thanks to Kelly for the extra support in all areas, especially when things were difficult.

Next, we come to the long list of friends and colleagues who have made my time in Bath so special. So, thank you to (in no particular order) Jon, Clarissa, Harry, Stephanos, Kerri, Tom, Rose, Kat, Rowan, James, Hasti and my bro's Ian, Jamie and Duncan. I wouldn't have had half as much fun as I've had if it wasn't for all of you, and I wish you all the best.

And finally, thank you to my amazing family for their undying love and support. To my grandparents, my brother Cerri, my sister-in-law Sam, my niece Lorelei, my mum and my dad. It is to you that I dedicate this thesis, I hope that you know how much I appreciate all that you have given me.

I love you all. And Dr. Love knows a thing or two about that.

Contents

Acronyms

AR	Anti-resonance/Anti-resonant
ARROW	Anti-resonant Reflecting Optical Waveguide
ASE	Amplified Spontaneous Emission
C	Cavity
CW	Continuous Wave
DC	Direct Current
FDP	Forward Double-Pass
FWHM	Full Width at Half Maximum
HCF	Hollow Core Fibre
ID	Inner Diameter
IR	Infrared
MMF	Multimode Fibre
OD	Outer Diameter
PBG	Photonic Bandgap Fibre
PCF	Photonic Crystal Fibre
RDP	Reverse Double-Pass
RF	Radio Frequency
SMF	Single Mode Fibre
SP	Single-Pass
TIR	Total Internal Reflection

Racah Spectroscopic Notation

There are numerous methods of defining the state of an excited atom. In this thesis, it was decided that Racah notation would be used for its ease in defining the states of neutral noble gases, and for its uniformity between different elements. As a combination of LS and J_1L_2 coupling, it defines both the state of the parent ion and the excited electron and is given by

$$\left({}^{(2S_1 + 1)}L_{1J_1} \right) nl[K]_J^o.$$

The initial bracketed term are the values of the total spin (S_1), orbital (L_1) and angular momentum (J_1) quantum numbers for the parent ion. The terms on the right define the excited electron, with principal (n) and orbital (l) quantum numbers and K and J the quantum numbers for $K = J_1 + l$ and $J = K + s$, respectively. The 'o' represents the parity of the excited atom. States that feature the symbol have an odd parity, whereas states without the symbol have an even parity.

For the neutral nobles gases, the parent ions can only be in the ${}^2P_{\frac{1}{2}}$ or ${}^2P_{\frac{3}{2}}$ states, and so the notation is often shortened to just the second term with a primed symbol over the orbital number to indicate the ${}^2P_{\frac{1}{2}}$ parent ion.

Chapter 1
Introduction

The research presented over the following 107/108 pages is a re-invention of the electrically pumped gas laser, dragging it kicking and screaming into the 21st century by replacing the cumbersome laser tube with new, state of the art hollow core optical fibres.

At its heart a laser consists of a light amplifying medium contained within an optical cavity [1, 2]. Atoms within the medium emit light which is trapped by the cavity and continuously amplified by subsequent emissions to a high intensity before it is released as a beam. The amplification requires a constant supply of energy to be pumped into the medium which keeps the atoms in their excited states; a process that can achieved electrically, optically or chemically.

Electrical pumping of a gaseous medium involves the production of a cloud of ionised atoms, also known as an electrical discharge [3]. Discharges can be generated through the direct application of a potential difference to the gas (DC pumping), or the indirect application of radio-frequency (RF) or microwave radiation. DC pumping is the simplest of these methods, requiring only a voltage source and resistance to provide current control. It is however also quite a dirty method with the electrodes being directly exposed to the gas, which can introduce impurities reducing the lifetime of the laser. RF excitation removes this issue but does introduce an extra layer of complexity in the electronics.

The first working electrically excited gas laser was demonstrated by Javan et al. at Bell labs in 1960 [4] using an RF-excitation method on a mixture of helium and neon in a 80 cm long tube with a 1.5 cm inner diameter to induce pulsed lasing on five neon emission lines around 1.15 μm. In the following years Patel et al. [5] reported a wide range of lines exhibiting gain in the noble gases including the 2.026 μm xenon line, while the first helium-xenon laser operating on the 3.508 μm xenon line was reported by Paananen and Bobroff in 1963 [6]. Both of these devices also employed an RF-excitation method to produce the discharge, with the earliest locatable report on a DC-excited device was in 1962 by Van Bueren et al. [7].

© Springer International Publishing AG, part of Springer Nature 2018
A. Love, *Hollow Core Optical Fibre Based Gas Discharge Laser Systems*,
Springer Theses, https://doi.org/10.1007/978-3-319-93970-4_1

As the density of a gaseous medium is that much less than that of a solid medium, there are significantly lower numbers of atoms available to emit light within gas lasers. This means that gas lasers typically require a large active volume, or a large pressure, to produce any appreciable power. Unfortunately, the electron dynamics in the discharges used to pump many gas lasers do not perform well in either high pressures or large diameter vessels. It has in fact been shown in the neutral noble gases that a smaller diameter discharge tube yields a higher gain [5, 8, 9]. This then leaves length as the only scaling factor to maximise volume. However as these early gas lasers relied on free-space Gaussian optics to confine their light, diffraction imposed a fundamental limit on both the maximum lengths and minimum diameters achievable [8].

In 1971 Smith attempted to overcome these limits by using sub-millimetre diameter tubes of up to 30 cm in length that would confine the light with the use of grazing incidence reflections on the glass surface [10]. Such waveguides still suffered significant losses scaling with $\frac{\lambda^2}{d^3}$ [11] and were extremely bend sensitive, but did bring some improvement.

Smith's first waveguide gas laser produced a continuous wave (CW) output power of 1 mW at 632 nm in a 10 : 1 ratio of ^{3}He : ^{20}Ne at a pressure of 7 Torr within a specially straightened 430 μm diameter capillary waveguide. This was followed up in 1973 by Smith and Maloney with a 5 : 1 mixture of ^{3}He : ^{136}Xe at a pressure of ~6 Torr in a 250 μm diameter capillary which exhibited gains of up to 800 dB m^{-1}, enough to produce significant frequency pulling effects [12]. Both of these devices used a combined DC and RF excitation method for stability.

In the forty years to follow no major advances were made to the fundamental gas laser design, and with the reputation of being large and fragile gas lasers had mostly fallen into disuse. However with the advent of low-loss hollow core optical fibres at the start of the 21st century [13] the diameter, length and straightness constraints imposed by diffraction can be removed leading to a fundamental change to the geometry of a gas laser as already demonstrated in optically pumped systems [14, 15]. Such a system could then be significantly longer to produce competitive powers, while also being cheaper and simpler than optically pumped systems and more compact than current gas lasers through the coiling of the fibre.

The generation of an electrical discharge within an optical fibre is not without its difficulties however; the reduced diameter and extended length increases the voltages required to achieve and sustain the discharge. Attempts at producing an RF or microwave pumped discharge laser have so far proved unsuccessful, with discharges of at most only 2 cm in length having been produced to date with no measurements of optical gain [16].

More success has been shown by DC excitation; in 2008 Shi et al. demonstrated discharges in fibres ranging from 2.9–26.2 cm in length and 50–250 μm in diameter [17]. These were however performed in high loss plain silica tubes, as the optically guiding hollow fibres of the time had core diameters of only ~10 μm and were too narrow to sustain the required plasma over long lengths. Fortunately though this did show that the required voltages for a discharge of this type are not excessive but in the order of ∼10 kV, a level that is more than attainable.

The development of low bend loss anti-resonant guiding hollow fibres with core sizes up to 100 μm at the University of Bath between 2012 and 2014 [18–20] combined with the work of Shi et al. [17] have finally made the dream of compact, flexible gas lasers an attainable goal.

This thesis describes the development of the first electrically excited gas lasers within a hollow optical fibre, based on DC discharges using mixtures of helium, neon and xenon gas. The content of this thesis is split into two parts; Chaps. 2–5 provide a more substantial introduction to the background physics on lasers, electrical discharges and optical fibres while Chaps. 6–8 describe the experimental work.

Chapter 6 covers the initial stages of the project including the construction of the vacuum system, electrical setup and optical cavity as well as early CW measurements. Chapter 7 covers all the experiments made with helium-xenon discharges measured in a pulsed regime. Chapter 8 covers similar experiments with helium, neon and xenon before concluding with a brief foray into argon.

References

1. J. Hecht, *Understanding Lasers: An Entry-Level Guide* (Wiley, 2008)
2. S.M. Hooker, C.E. Webb, *Laser Physics* (Oxford University Press, 2010)
3. C.S. Willett, *Introduction to Gas Lasers: Population Inversion Mechanisms* (Pergamon Press, 1974)
4. A. Javan, W.R. Bennett, D.R. Herriott, Population inversion and continuous optical maser oscillation in a gas discharge containing a He–Ne mixture. Phys. Rev. Lett. **6**(3), 106 (1961)
5. C. Patel, W. Bennett, W. Faust, R. McFarlane, Infrared spectroscopy using stimulated emission techniques. Phys. Rev. Lett. **9**(3), 102 (1962)
6. R.A. Paananen, D.L. Bobroff, Very high gain gaseous (Xe–He) optical maser at 3.5 μm. Appl. Phys. Lett. **2**(5), 99 (1963)
7. H. Van Bueren, J. Haisma, H. De Lang, A small and stable continuous gas laser. Phys. Lett. **2**(7), 340 (1962)
8. P.W. Smith, On the optimum geometry of a 6328Å laser oscillator. IEEE J. Quan. Electr. **2**(4), 77 (1966)
9. P.O. Clark, Investigation of the operating characteristics of the 3.5 μm xenon laser. IEEE J. Quant. Electr. **1**(3), 109 (1965)
10. P.W. Smith, A waveguide gas laser. Appl. Phys. Lett. **19**, 132 (1971)
11. E.A.J. Marcatili, R.A. Schmeltzer, Hollow metallic and dielectric waveguides for long distance optical transmission and lasers. Bell Syst. Tech. J. **43**(4), 1783 (1964)
12. P.W. Smith, P.J. Maloney, A self-stabilized 3.5 μm waveuide He–Xe laser. Appl. Phys. Lett. **22**, 667 (1973)
13. R.F. Cregan, Single-mode photonic band gap guidance of light in air. Science **285**(5433), 1537 (1999)
14. A.M. Jones, A.V.V. Nampoothiri, A. Ratanavis, T. Fiedler, N.V. Wheeler, F. Couny, R. Kadel, F. Benabid, B.R. Washburn, K.L. Corwin, W. Rudolph, Mid-infrared gas filled photonic crystal fiber laser based on population inversion. Opt. Exp. **19**(3), 2309 (2011)
15. M.R.A. Hassan, F. Yu, W.J. Wadsworth, J.C. Knight, Cavity-based mid-IR fiber gas laser pumped by a diode laser. Optica **3**(3), 218 (2016)

16. F. Vial, K. Gadonna, B. Debord, F. Delahaye, F. Amrani, O. Leroy, F. Gérôme, F. Benabid, Generation of surface-wave microwave microplasmas in hollow-core photonic crystal fiber based on a split-ring resonator. Opt. Lett. **41**(10), 2286 (2016)
17. X. Shi, X.B. Wang, W. Jin, M.S. Demokan, Investigation of glow discharge of gas in hollow-core fibers. Appl. Phys. B Lasers Opt. **91**(2), 377 (2008)
18. F. Yu, W.J. Wadsworth, J.C. Knight, Low loss silica hollow core fibers for 3–4 μm spectral region. Opt. Exp. **20**(10), 11153 (2012)
19. W. Belardi, J.C. Knight, Hollow antiresonant fibers with low bending loss. Opt. Exp. **22**(8), 10091 (2014)
20. F. Yu, J.C. Knight, Negative curvature hollow-core optical fiber. IEEE J. Sel. Top. Quant. Electr. **22**(2) (2016)

Chapter 2
Introduction to Laser Physics

The laser (**l**ight **a**mplification by **s**timulated **e**mission of **r**adiation), first demonstrated by Theodore Maiman in 1960 [1], has found considerable uses ranging from
manufacturing, spectroscopy, telecommunications, metrology and even recreational,
all thanks to its nature of producing highly monochromatic, spatially coherent light.

This chapter introduces some basic concepts of laser oscillation that will be discussed later in this thesis. The topics that will be covered include; the most basic
picture of a laser, the requirements of a gain medium, the definition of optical gain,
the behaviour of a laser resonator and the nature of resonator modes.

There is an extensive supply of textbooks available that cover the material presented below, though references [2, 3] provide a good entry level introduction to the
subject.

2.1 Types of Atomic Transition

An atom (or ion) in a two energy level system can interact with a photon in one of three
ways as illustrated in Fig. 2.1. If said atom is in the lower state it can *absorb* a photon
of the correct energy and make the transition to the upper energy level. Conversely, if
said atom is in the upper state it can randomly emit a photon of corresponding energy
and make the transition to the lower energy level, a process know as *spontaneous
emission*.

In the third case a photon of the correct energy induces an excited atom to emit an
photon identical to the incoming photon. This process is know as *stimulated emission*,
and is the most fundamental principle of a laser.

© Springer International Publishing AG, part of Springer Nature 2018
A. Love, *Hollow Core Optical Fibre Based Gas Discharge Laser Systems*,
Springer Theses, https://doi.org/10.1007/978-3-319-93970-4_2

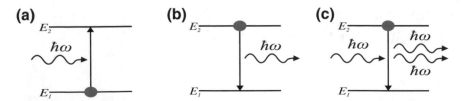

Fig. 2.1 Simple energy level diagrams for **a** absorption, **b** spontaneous emission and **c** stimulated emission, all involving a photon of energy $E = E_2 - E_1 = \hbar\omega$

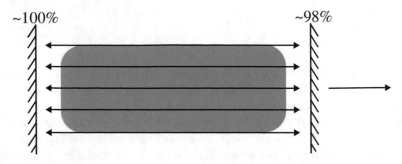

Fig. 2.2 Schematic diagram of a simple laser. The shaded area is representative of a gain medium. Light is trapped within the medium by two mirrors on either side. One of the mirrors has a slightly lower reflectance ($\sim$98% here) to release the laser beam, while the other is as reflective as possible (ideally 100%)

2.2 A Laser at Its Simplest

The simplest picture of a laser requires just two components; a gain medium where stimulated emission can occur and a cavity to 'trap' the light (see Fig. 2.2). The cavity contains light emitted by the medium, which in turn triggers more and more stimulated emission to produce a high intensity of strongly coherent light. A fraction of this light is released from one end of the cavity as the output laser beam.

2.3 Population Inversion

The picture presented in Sect. 2.2 will only work if there is consistently a higher population of atoms in the upper excited state than in the lower state, otherwise more incoming photons will be absorbed than stimulate an emission. However, for a system that is in thermodynamic equilibrium it is the opposite that will always be true. In order to achieve such a *population inversion*, the medium requires *pumping* from an external source of energy. This pumping can be achieved electrically, optically or chemically.

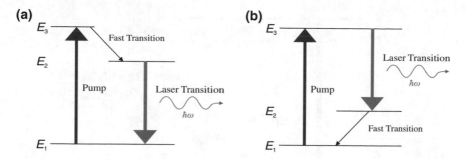

Fig. 2.3 Energy diagrams of 3-level laser systems as used in **a** the first ruby laser and **b** electrically pumped gas lasers

A population inversion can be boosted in a number of different ways. One of these is through the exploitation of metastable energy levels; states that an atom can exist in for time scales up to micro- or even milliseconds before spontaneously emitting as opposed to the nanosecond lifetimes of 'standard' states. Another is to involve additional fast decaying states within the transition cycle.

Most real laser systems rely on a large number of energy levels. These can usually be simplified to being either a 3- or 4-level system. Two examples of 3-level systems can be seen in Fig. 2.3. The system shown in Fig. 2.3a is representative of the system used in the first optically pumped ruby lasers [1].

With an equal probability between absorption and stimulated emission, optical pumping can at best only bring two levels to equilibrium. The addition of a third level acting as the upper lasing state which is connected to the pumped level via a fast, non-radiative transition enables an inversion to be built up between it and the ground. However, because this system uses the ground state as the lower lasing level, more pumping is required to depopulate the ground state and reduce absorption. This limited the ruby laser to operating only in the pulsed regime.

Figure 2.3b is of an alternative scheme, utilised by many electrically pumped gas lasers. With a non-optical pump a limited inversion through direct excitation can be achieved thanks to radiation trapping amongst other phenomena (see Chap. 4). However without additional levels achieving a sizeable inversion is still difficult. Using an extra level as a lower lasing level connected to the ground via a fast transition means that it will be quickly depleted, and even a small population in the upper state can provide a considerable inversion.

Many lasers will use a combination of these two systems. By separating both the upper and lower lasing levels from the pumped and ground levels as in a 4-level system (as in Fig. 2.4), the advantages of the previous two systems can be exploited without the disadvantages. Lasers that use this system include the Nd:YAG laser, and IR gas recombination lasers [4].

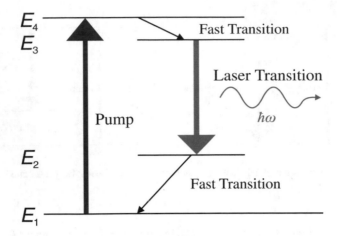

Fig. 2.4 Energy diagram of a 4-level laser system

2.4 Optical Gain and the Gain Bandwidth

Optical gain is a measure of the amplification ability of a medium, in other words
it is the amount of stimulated emission that a photon travelling through the medium
can trigger. The gain of a medium is usually expressed by its small signal gain α_0,
and its amplification factor

$$A = \exp(\alpha_0 l), \tag{2.1}$$

where l is the length of the gain medium.

From the energy diagrams in Fig. 2.3 it can be easy to believe that atomic emission
and subsequent gain is exactly monochromatic, but in reality any given transition has
a defined *bandwidth* inside of which an emission can occur. Its existence is a result
of the uncertainty principle, which relates the lifetime of a state to an uncertainty in
its energy.

Collisions between particles can cause this natural bandwidth to be further broadened as they reduce the lifetimes of the states, increasing the uncertainty in the
location of an energy level. This broadening will subsequently be proportional to the
rate of the collisions, so increasing the pressure of a gas laser medium will increase
the bandwidth. This mechanism is known as *pressure* or *collisional broadening* and
results in a bandwidth that is Lorentzian in shape [3].

Another broadening method comes from the constant erratic movement of gaseous
atoms. As each atom moves independently to one another they all experience their
own Doppler shift. When the emission frequencies of all these individual shifted
atoms are summed together the emission line broadens to a Gaussian. This *Doppler
broadening* is given by [3]

$$\Delta f = f_0 \left(\frac{8k_B T \ln(2)}{Mc^2} \right)^{\frac{1}{2}},$$ (2.2)

where Δf is the broadening around an emission line of frequency f_0 from an atom of mass M, c is the speed of light, k_B the Boltzmann constant and T the absolute temperature.

As the bandwidth of each individual atom in a Doppler broadened medium is smaller than the overall gain bandwidth, an emission from one atom may not necessarily be able to stimulate an emission in a second atom with a different velocity. Doppler broadening is hence referred to as an *inhomogeneous* broadening. Conversely *homogeneous* broadening methods such as pressure broadening affects each atom in the same way, so the bandwidth of each atom matches the overall gain bandwidth.

2.5 Amplified Spontaneous Emission

Atoms within an excited gain medium will emit radiation in all directions. If a photon is spontaneously emitted so that it travels through the medium it has the potential to stimulate many more emissions, resulting in a phenomenon known as *Amplified Spontaneous Emission* or ASE. The effect is particularly prevalent if an atom at one end of a medium confined within a waveguide emits a photon such that it propagates the length of the waveguide.

In mediums that exhibit very high gain, it is possible to reach saturation from ASE alone (see Sect. 2.7). Such cases are not widely regarded as a true laser, but are instead referred to as *mirror-less lasers* [3].

2.6 Laser Cavity Modes

As with any cavity, interference between the rebounding waves within a laser cavity will set up longitudinal modes along the length of the cavity. These modes will occur at frequencies [2]

$$\nu_N = \frac{N}{T_C},$$ (2.3)

where N is an integer and $T_C = \frac{2nL}{c}$ is the cavity round trip time for a cavity of length L and refractive index n. Any modes that fall within the gain bandwidth will experience amplification (see Fig. 2.5).

The presence of multiple longitudinal modes, each with a minor variation in frequency, results in a beating between the modes. This beating is exhibited in the

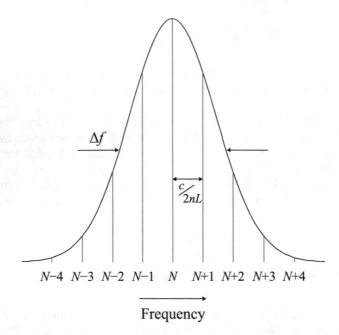

$$N-4 \quad N-3 \quad N-2 \quad N-1 \quad N \quad N+1 \quad N+2 \quad N+3 \quad N+4$$

Frequency

Fig. 2.5 An illustration of the gain bandwidth of an emission line (think line) and the longitudinal modes contained within (thin lines)

temporal domain by oscillations in output power with a frequency equal to the mode separation.

2.7 The Lasing Threshold and Gain Saturation

For a medium contained within a cavity, lasing can occur in a particular mode only if the round trip amplification matches the round trip loss. This can be formally expressed by the *threshold condition*

$$R_1 R_2 \exp(2\alpha_0^{th} l) \exp(-2\kappa_L L) = 1, \tag{2.4}$$

where $R_{1,2}$ are the cavity mirror reflectances, α_0^{th} the threshold gain and κ_L the cavity loss. Above this threshold the gain is higher than the cavity loss, hence the circulating power builds until the excited atoms are stimulated to emit just as fast as they are pumped to their upper states. The gain is said to be *saturated*, and is fixed to the threshold value for the particular longitudinal mode that meets this condition. The small signal gain α_0 for the oscillating mode is replaced by the saturated gain α_1. Any further increases in pumping power will only increase the circulating and output powers, leaving the gain unchanged (see Fig. 2.6).

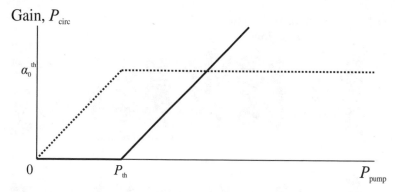

Fig. 2.6 Cavity gain (dotted) and circulating power (solid) as a function of pump power

2.8 Multimode Lasing

The finer points of laser oscillation depend on the broadening method of the gain medium. As the bandwidth of each individual atom matches the overall bandwidth in a homogeneously broadened medium, as soon as one mode begins to lase the gain is saturated across the whole bandwidth and becomes fixed. This means that only the mode closest to the centre of the gain bandwidth could ever be able to oscillate, and the laser should only operate on a single longitudinal mode.

However due to the nature of standing waves in a cavity, and the existence of nodes and antinodes, multimoded operation of homogeneously broadened lasers is possible in some cases. This is demonstrated in Fig. 2.7, where the central lasing mode (a) has saturated the gain at the high intensity antinodes, but left the gain unsaturated at the low intensity nodes. The presence of unsaturated regions in the medium pave the way for mode (b), of a different frequency with antinodes at different positions, to pick up enough gain to reach the oscillation threshold. This non-uniform 'burning' of the gain medium is known as *spatial hole-burning*.

In an inhomogeneously broadened laser however, where the total bandwidth is the sum of each atom's individual bandwidth, once the first mode begins to oscillate it is only a narrow range of frequencies which become saturated (see Fig. 2.8a). Subsequent increases in pump power will increase the gain for all other frequencies while the gain for the oscillating mode will be fixed at the threshold, creating a *spectral hole* in the saturated gain coefficient (see Fig. 2.8b). Once the gain at other available modes reaches threshold, that mode is free to lase with the first, drilling its own spectral hole in α_1, thus simultaneous oscillation on multiple modes is possible (see Fig. 2.8c).

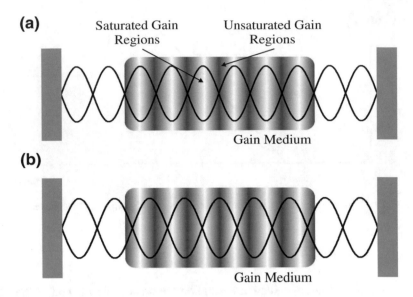

Fig. 2.7 Spatial hole-burning in a homogeneously broadened laser, demonstrated by two cavity modes of order **a** $N = 10$ and **b** $N = 9$. The antinodes of mode (a) deplete the gain of the medium in specific regions, indicated by the white bands, leaving regions of high gain for the antinodes of mode (b) to feed off of

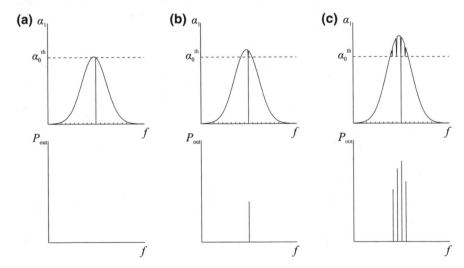

Fig. 2.8 Saturated gain coefficient α_1 and output power P_{out} for different longitudinal modes in an inhomogeneously broadened laser as pumping is increased (a) to threshold (b) just above threshold and (c) further above threshold

2.9 Frequency Pulling

As the real part of the refractive index has a slight dependence on gain, the mode frequency calculation in Eq. 2.3 is strictly speaking only valid in the absence of a population inversion, or a *cold cavity*. Within a *hot cavity*, where an inversion is present, Eq. 2.3 becomes

$$v'_N T_C + v'_N \frac{2 \Delta n^r_g l}{c} = N, \tag{2.5}$$

where v'_N the frequency of mode N in a hot cavity and the real part of the refractive index shift due a homogeneously broadened gain line of width Δf centred at f_0 is given by [3]

$$\Delta n^r_g = \frac{2 \pi \alpha_1 c}{v_N} \frac{(v_N - f_0)}{\Delta f}. \tag{2.6}$$

By combining Eqs. 2.3, 2.5 and 2.6, the frequency of mode N in a hot cavity can be expressed as

$$v'_N = \frac{\left(\frac{T_C}{4 \pi \alpha_1 l}\right) v_N + \left(\frac{1}{\Delta f}\right) f_0}{\left(\frac{T_C}{4 \pi \alpha_1 l}\right) + \left(\frac{1}{\Delta f}\right)}, \tag{2.7}$$

and so the frequency spacing between mode N and $N + 1$ can be written

$$\Delta v' = \frac{\Delta v}{1 + \frac{4 \pi \alpha_1 l}{T_C \Delta f}}. \tag{2.8}$$

From Eq. 2.8 it is apparent that as the gain increases, the mode frequency spacing is decreased - the frequencies of the hot cavity modes are 'pulled' towards the centre frequency. In general this frequency pulling is $\sim 10^3$ smaller than the mode frequencies and can hence be neglected unless the medium exhibits a particularly high gain and a bandwidth that is narrow compared to the cold cavity mode spacing ($\alpha_1 \sim T_C \Delta f$).

References

1. T.H. Maiman, Stimulated optical radiation in ruby. Nature **187**(4736), 493 (1960)
2. J. Hecht, *Understanding Lasers: An Entry-Level Guide*, Wiley (2008)
3. S. M. Hooker, C. E. Webb, *Laser Physics*, Oxford University Press (2010)
4. W.J. Witteman, P.J.M. Peters, H. Botma, Y.B. Udalov, IR recombination lasers. Infrared Phys. Technol. **36**(1), 529 (1995)

Chapter 3
Introduction to Discharge Physics

The basis of an electrically pumped gas laser is the gas discharge. This chapter covers the basics of direct current (DC) discharges including how breakdown is achieved, the structure of a discharge, the key parameters involved, the particle dynamics within the discharge and methods of operating a pulsed discharge. Further detail of the material covered here and more can be found in [1, 2].

3.1 What is a Gas Discharge?

When a potential difference is applied to a low pressure gas between two electrodes, electrons emitted from the negative cathode can collide with the neutral gas atoms or molecules, ionising some of them. The resulting cloud of partially ionised gas particles is known as a *gas discharge*.

A simple circuit to create gas discharges can be seen in Fig. 3.1. All that is required is a container for the gas and electrodes, a voltage supply and a ballast resistance to regulate the current in the circuit. The three important parameters here are the supply voltage V_s, the voltage across the discharge V_d and the current I.

3.2 Modes of Operation

A typical gas discharge has a number of different modes of operation which are defined by the current and voltage as seen in Fig. 3.2. When a small potential difference is applied the discharge is in the *dark current mode* as at this point the discharge is not visible to the eye. In this stage a small amount of ionisation from background radiation sources is present, and the resulting electrons and ions are swept out by the electric field to give a very small current.

Increasing the supply voltage increases this current very slowly until a specific breakdown voltage V_{br} is reached, indicated in Fig. 3.2 by the dashed line. At this

© Springer International Publishing AG, part of Springer Nature 2018
A. Love, *Hollow Core Optical Fibre Based Gas Discharge Laser Systems*,
Springer Theses, https://doi.org/10.1007/978-3-319-93970-4_3

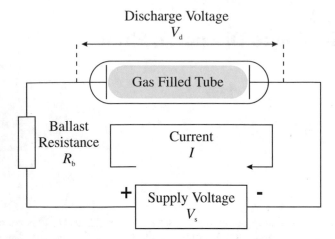

Fig. 3.1 A simple gas discharge circuit

point the discharge voltage drops and the system makes the transition into the *glow mode*. V_{br} is dependent on a number of factors; the type of gas, the shape and material of the electrodes, the pressure and the spacial parameters of the containment vessel, and is described by Paschen's Law [2]. The transition between the dark and glow modes is hysteretic in nature, with the current and supply voltage required to sustain the glow discharge lower than they are immediately after breakdown (as shown in Fig. 3.2 by the two dotted lines at the transition).

The glow discharge is the most commonly used type, so named for its characteristic glow. Used in many types of lighting, it is this mode that forms the basis of gas discharge laser systems. For this reason it will be the primary subject in the following sections.

The glow mode itself can then be split into two further sub-modes. At lower currents, the discharge is in the *normal* glow mode, where the discharge has negative differential resistance. This means that as the current is increased, the potential difference across the discharge decreases. This behaviour can be exploited for use in oscillation circuits such as the dynatron oscillator [3, 4].

If the current continues to increase, the discharge passes into the *abnormal* glow mode, where it re-establishes a positive differential resistance. Further increases in current push the discharge into the *arc mode*. At these currents the cathode can reach sufficient temperatures to emit electrons thermionically, hence thermal conditions have a much greater effect on the discharge behaviour. An arc discharge differs from a glow discharge in that the ion temperature is roughly equivalent to the electron temperature (see Sect. 3.4.2), and the emission spectrum includes lines from the cathode material.

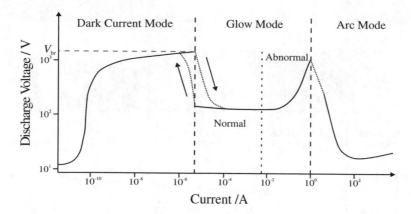

Fig. 3.2 A voltage-current plot for a typical gas discharge, showing the three main modes of operation. The fast transitions between the modes are highlighted by dotted lines. Axis scales are representative of a general case, as the real electrical parameters are dependent on many factors including the vessel dimensions and gases present

3.3 Structure of a Glow Discharge

A glow discharge comprises of bands of luminous and dark regions as seen in Fig. 3.3. Starting from the cathode, the first region is the Aston dark space which is where the electric field is at its strongest. At this point the electrons emitted from the cathode, known as *primary* electrons, do not possess sufficient energy to excite the gas particles so the region is dark, but the electrons are accelerated towards the anode.

At the point where the average electron kinetic energy is sufficient to excite the neutral gas particles and cause photon emission, a small luminous band appears known as the *cathode glow*. As the electric field is still strong here, the electrons are accelerated further. After the cathode glow layer, a dark region of little to no light known as the cathode dark space exists where the kinetic energy of the electrons is greater than the maximum excitation energy of the neutral gas particles. Collectively, the Aston dark space, cathode layer and cathode dark space can be referred to as the cathode fall space.

Following the cathode dark space is the first of the strong bright regions of the glow discharge known as the *negative glow* (indicted by the red region in Fig. 3.3). The negative glow region is maintained by the few remaining high energy primary electrons that collide with and ionise the neutral gas particles. Having lost energy in such collisions, the primary electrons have passed below the maximum excitation energy threshold and are free to cause the extensive excitations that are responsible for the intense glow of the region.

As the negative glow is traversed, the number of ionisation and excitation events that have occurred increase, creating more electron-ion pairs. These electrons freed by collisional ionisation are referred to as *secondary* electrons. The concentration of secondary electrons and ions in the centre of negative glow can be up to 20

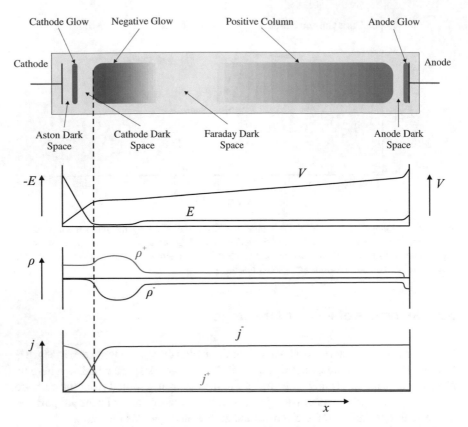

Fig. 3.3 Distribution of dark and luminous regions, electric field E, accumulated potential difference from the cathode V, positive and negative space-charge densities ρ^+ & ρ^- and current densities j^+ & j^- for the ions and electrons respectively in a glow discharge

times greater than those of the other regions [5]. As these electrons and ions start to recombine deeper into the negative glow it begins to fade to make way for the *Faraday dark space*. The electrons within this region again have insufficient energy to cause excitations, and are instead accelerated by the weak electric field.

When the energy of the electrons is once again sufficient to excite the neutral gas particles, the second most luminous region of the discharge known as the *positive column* begins (indicated by the blue region in Fig. 3.3). Unlike the negative glow, the stability of this region is determined by the diffusion of the ionised gas and so is affected by the radial dimensions of the containment vessel. For a stable system, the ionisation rate needs to match the rate of electron and ion loss from recombination and diffusion.

Although similar in appearance, the negative and positive regions of the discharge behave very differently as the dimensions of the discharge housing are altered. While increasing (or decreasing) the size of the vessel (both length and width) will have no real effect on the negative glow, the positive column will extend (or contract) to fill the remaining space. The negative glow however behaves as if it were 'stuck' to the cathode and can be moved by rotating the cathode while the positive column again just fills the space between the Faraday dark space and the anode. These behaviours suggest that the negative glow features very 'beam-like' particle dynamics, while the positive column is dominated by much more random dynamics.

Due to electron attraction and ion repulsion by the anode, a negative space charge is established around the anode. The resulting field that arises from this space charge accelerates electrons towards the anode, so low energy electrons exiting the positive column into the anode dark space may once again have enough energy to cause some final excitations before reaching the anode. Those excitations result in a luminous sheath around the anode called the *anode glow*.

3.4 Theory of the Positive Column

In electrically pumped gas lasers, it is the positive column that makes up the bulk of the discharge and so it is in this region that the most important dynamics for the operation of a laser take place. Before these can be described, a number of key parameters first need to be defined.

3.4.1 Particle Collisions and the Mean Free Path

Collisions between the atoms, ions and electrons are the driving force behind all the physical effects present in the positive column. These collisions can be categorised as elastic, inelastic or superelastic.

An *elastic* collision is where the total kinetic energy of two colliding bodies is conserved. Within a gas discharge system, the collisions between fast, light electrons and slow, heavier gas particles that cause no excitation are elastic.

Inelastic collisions are between two bodies where the total kinetic energy is not conserved, and decreases. Within a gas discharge system these include any collision that causes an excitation or ionisation, whether it between an electron and gas particle, or two gas particles.

Superelastic collisions are between two bodies where the total kinetic energy is not conserved, and increases. Within a gas discharge system they occur when an electron collides with an excited neutral gas particle and causes a de-excitation with the energy being transferred to the electron.

An important parameter when considering the collisional dynamics of particles is the *mean free path*, defined as the average distance a particle travels between collisions. Assuming that the particles behave as hard spheres, the mean free path of a type 1 particle present in a gas consisting of N type 2 particles m^{-3} is given by

$$\lambda_1 = \frac{1}{N\pi(r_1 + r_2)^2\left(1 + \frac{v_2^2}{v_1^2}\right)^{\frac{1}{2}}}, \tag{3.1}$$

where $r_{1,2}$ and $v_{1,2}$ are the radii and velocities of particle types 1 and 2 respectively
[1]. For collisions between atoms of equal size $r_1 = r_2 = r_a/2$ and $v_1 \approx v_2$, and
for electrons moving in gases $r_1 = r_e << r_2$ and $v_1 = v_e >> v_2$, giving respective
mean free paths of

$$\lambda_a = \frac{1}{N\pi r_a^2 \sqrt{2}} = \frac{1}{N\sigma_a} \tag{3.2}$$

$$\lambda_e = \frac{1}{N\pi r_e^2} = \frac{1}{N\sigma_e}, \tag{3.3}$$

where $\sigma_{a,e}$ are the collisional cross-sections for the atoms and electrons respectively.
The cross-section is a way of representing the probability of a collisional event. It is
a hypothetical area around a particle which indicates that a collision has occurred if
crossed by another incoming particle.

3.4.2 The Electron Temperature

The *electron temperature* of a plasma is equivalent to the average kinetic energy
of the free electrons within the plasma. It is dependent on the species of ions, the
applied field, the pressure and the containment vessel dimensions. For a large enough
electron population ($>10^{16}$ m^{-3}) in thermodynamic equilibrium with each other, the
electron temperature T_e can be defined by

$$\frac{m_e \langle v_e^2 \rangle}{2} = \frac{3k_B T_e}{2}, \tag{3.4}$$

where $\langle v_e^2 \rangle$ is the mean square velocity of the free electrons that are defined by a
Maxwellian distribution, m_e is the electron mass and k_B is the Boltzmann constant.

3.4.3 The Debye Length

The *Debye length* is defined as the length scale at which the positive charge of an
ion in a plasma can be felt before it is 'screened' out by the surrounding cloud of
free electrons. At distances from the ion greater than this length scale a plasma can
be treated as quasi-neutral. The Debye length is given by [6]

$$\Lambda_D = \left(\frac{k_B T_e \epsilon_0}{e^2 N_e}\right)^{\frac{1}{2}} \approx \left(\frac{T_e}{N_e}\right)^{\frac{1}{2}} \times 70 \text{ m}^{-1} \text{ K}^{-1}, \tag{3.5}$$

where ϵ_0 is the permittivity of free space, e is the electron charge and N_e is the density of electrons. In a typical gas discharge T_e can vary from 10^4 to 10^5 K and N_e from 10^{16} to 10^{20}, giving a theoretical range of Λ_D from 0.7 μm to 221 μm.

3.4.4 The Positive Column in a Large Radius, Cylindrical Container

In the initial stages of an electrical discharge, fast moving electrons become attached to the walls of the discharge tube, giving the walls a slight negative potential and setting up a radial electric field. Provided that the radius of the tube R is much greater than the electron and ion mean free paths λ_e and λ_i and the Debye length Λ_D (so that there will be more collisions amongst the particles than with the tube walls and charge neutrality can be assumed), then the radial field will cause the electrons and ions to diffuse outwards together. This process is known as *ambipolar diffusion*.

Equating the number of charge pairs N (where $N = N_e = N_i$ through charge neutrality) lost by diffusion to the walls with the number of pairs gained by ionisation within a tubular volume element of thickness dr per unit length of the cylinder yields the differential equation

$$\frac{d^2 N}{dr^2} + \frac{1}{r}\frac{dN}{dr} + \frac{\alpha}{D_a}N = 0, \tag{3.6}$$

where D_a is the coefficient of ambipolar diffusion and α is the rate of ionisation per electron, which for a Maxwellian distribution of electrons of concentration n_e in a gas whose ionisation potential is V_i are given as

$$D_a \approx \frac{b^+ k_B T_e}{e} \tag{3.7}$$

$$\alpha = 600\left(\tfrac{2}{\pi}\right)^{\frac{1}{2}} ap\left(\frac{e}{m_e}\right)^{\frac{1}{2}} V_i^{\frac{3}{2}} x^{-\frac{1}{2}} \exp(-x), \tag{3.8}$$

with

$$x = \frac{eV_i}{k_B T_e}, \tag{3.9}$$

where b^+ is the ion mobility, p the pressure and a the efficiency of ionisation (see [2] for the full derivation). Equation 3.8 has been plotted in Fig. 3.4 as a function of electron temperature for clarity.

The solution to Eq. 3.6 is a Bessel function of zero order which has a first zero at 2.405. Assuming that ion recombination occurs at the wall only, the boundary conditions of $n_e = 0$ at $r = R$ can be applied yielding

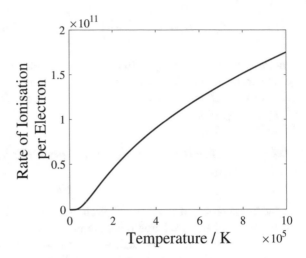

Fig. 3.4 Rate of ionisation per electron in a xenon gas at a pressure of 9 torr as a function of electron temperature

$$\frac{\alpha}{b^+} = \frac{k_B T_e}{e} \left(\frac{2.405}{R}\right)^2. \tag{3.10}$$

Inserting α in Eq. 3.10 and rearranging then gives

$$\left(\frac{k_B T_e}{e V_i}\right)^{\frac{1}{2}} \exp\left(\frac{e V_i}{k_B T_e}\right) = 1.2 \times 10^7 (c^* p R)^2, \tag{3.11}$$

where

$$c^* = \left(\frac{a V_i^{\frac{1}{2}}}{b^+ p}\right)^{\frac{1}{2}}. \tag{3.12}$$

Equation 3.11 is plotted in Fig. 3.5, where it is clear that as the product of the gas pressure p and tube radius R - or more conveniently tube diameter D - decreases, the electron temperature increases. This is because more ions are lost to the tube walls, and so the electron temperature must rise to compensate for this.

Higher electron temperatures can be potentially problematic in smaller diameter tubes due to its proportionality with the Debye length. If T_e is high enough so that Λ_D is of the same order as the discharge tube radius then instabilities can arise. However thanks to the dependence of T_e on the pD product, this can be controlled by increasing p to compensate.

Below 2 Torr-mm (2.7 mbar-mm) Eq. 3.11 is not strictly valid as the space charge 'sheath' set up by the slight negative potential on the discharge tube walls becomes comparable in size to the tube radius. As the subject of small radius discharge tubes is important in this context of this thesis it is discussed further in Sect. 3.4.5.

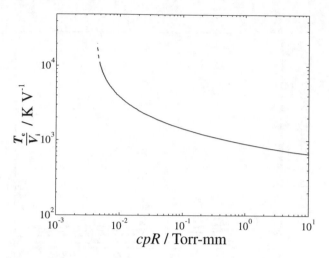

Fig. 3.5 Normalised electron temperature $\frac{T_e}{V_i}$ as a function of $c^* p R$. Typical c^* values for the noble gases are; helium - 4×10^{-3}, neon - 6×10^{-3}, argon - 4×10^{-2} (values taken from [2])

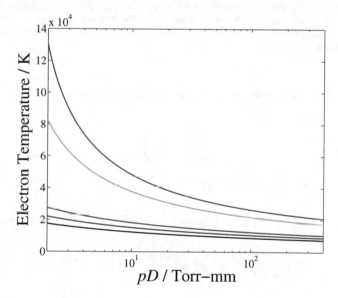

Fig. 3.6 Electron temperature T_e as a function of $p D$ for helium (red), neon (cyan), argon (blue), krypton (magenta) and xenon (black) [1]

Figure 3.6 is a replica of Fig. 3.5 but for the noble gases only. As argon, krypton and xenon all have low ionisation potentials and mobilities, their discharges are maintained at much lower electron temperatures than for helium and neon.

While the preceding analysis is only applicable to a single gas, it can be adapted to calculate the electron temperature in a two gas mixture. This adaptation involves

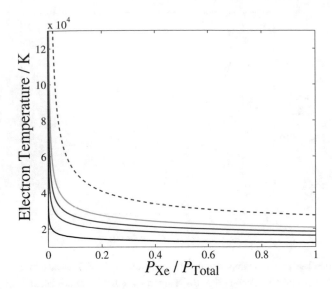

Fig. 3.7 Electron temperature T_e as a function of the partial pressure of Xe in a mixture of He-Xe for pD values of 0.9 Torr-mm (1.2 mbar-mm) (dotted red), 2 Torr-mm (2.7 mbar-mm) (cyan), 3 Torr-mm (4.0 mbar-mm) (blue), 5 Torr-mm (6.7 mbar-mm) (magenta) and 20 Torr-mm (26 mbar-mm) (black) based on Eq. 3.13 [1]

replacing the single ionisation rate and ion mobility terms α and b^+ with the combined rates of

$$\left(\frac{\alpha_1}{b_1^+} + \frac{\alpha_2}{b_2^+}\right).$$

Eqs. 3.8, 3.10 and 3.11 then become [7]

$$(pD)^2 \left(f_1 c_1^{*2} x_1^{-\frac{1}{2}} \left[1 + \frac{1}{2}x_1\right] \exp(-x_1) + f_2 c_2^{*2} x_2^{-\frac{1}{2}} \left[1 + \frac{1}{2}x_2\right] \exp(-x_2)\right) =$$
$$= \left[\left(\frac{2}{300\pi}\right)^{\frac{1}{2}} \left(\frac{1}{2.405}\right)^2 \left(\frac{e}{m}\right)^{\frac{1}{2}}\right]^{-1} \left(= 1.72 \times 10^{-7} \, \mathrm{V}^{\frac{1}{2}} \, \mathrm{s\,cm}^{-1}\right), \quad (3.13)$$

where $c_{1,2}^*$, $x_{1,2}$ are defined as before but for each of gases 1 and 2, $f_{1,2}$ are the ratios of partial pressures to total pressure and $\frac{e}{m}$ is given in esu units (the factor 300 comes from the conversion from esu potential to volts) as is given in [7]. Figure 3.7 illustrates how T_e depends on the partial pressure of xenon in a He-Xe mixture for different values of pD. Again, due to the high ionisation potential of helium, T_e increases as helium is added to the mixture.

This argument however only includes ionisation from electron collisions, which in He-Ar, He-Kr or He-Xe mixtures is not strictly true. As the metastable states of helium are at higher energies than the ionisation energy of the second gas (Ar, Kr

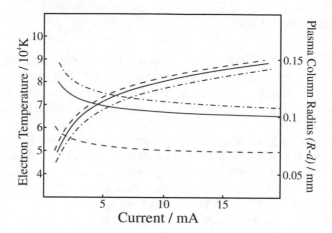

Fig. 3.8 Calculated dependence of electron temperature (blue) and plasma column radius (red) on discharge current for pure He (dashed), pure Ne (dotted) and an 8:1 mixture of He-Ne (solid) all at a pressure of 10 Torr (13.3 mbar) in a discharge tube of radius 0.2 mm. Recreated from [8]

or Xe), collisions with an excited helium atom can cause the ionisation of an atom from the other specie. A collisional atom-atom ionisation such as this is referred to as *Penning Ionisation*, the general representation of which is given by

$$A^* + B \rightarrow A + B^{+(*)} + e^-, \tag{3.14}$$

with atom A being the initially excited atom, and B being a second atom that has ended up in one of a number of excited ionised sates.

3.4.5 Small Radius Behaviour

As previously mentioned, at low values of pD the space charge sheath surrounding the discharge tube walls becomes comparable in size to the tube radius, and so has a much bigger impact on the dynamics of the positive column. The central plasma region behaves as before, only it is now restricted further by the space charge sheath. Within the sheath ions are accelerated toward the negative potential of the wall, while electrons are decelerated or reflected. The rate of ionisation within the sheath is subsequently reduced to the point where it can be neglected.

A full description is offered in reference [8] where it is shown that the radial extent of the sheath d is not only strongly dependent on R and p, but also on the discharge current I. Figure 3.8 shows how the central plasma region radius $R - d$ and hence the electron temperature depends on the discharge current. It is clear that as the current is decreased, T_e increases, and so it is harder to maintain a stable discharge at lower currents where operation of a laser is at its maximum efficiency [9, 10].

3.5 Pulsed Discharges

There are a number of methods that can be employed to produce a pulsing discharge, ranging from simply switching the power supply on and off again to more elegant methods of constructing oscillatory circuits. The simplest of these is the *Pearson-Anson oscillator* [11], where a capacitor is connected in parallel to the discharge vessel as in Fig. 3.9a.

The voltage-time graph in Fig. 3.9b demonstrates how the circuit operates. Once the supply voltage is turned on the capacitor starts to charge (1). This continues until the voltage across the capacitor reaches V_{br} where the discharge strikes, allowing a current to flow though it hence discharging the capacitor (2). The voltage across the capacitor then drops rapidly, but because of the hysteretic nature of the glow discharge it doesn't extinguish until the voltage drops below the extinction voltage V_{ex} (3). Once extinguished, the capacitor starts charging again and the process repeats itself (4), resulting in the sawtooth-like voltage profile in Fig. 3.9b.

In most cases, the time that the discharge is on is much less than the time than it is off, and so the characteristic time period of pulsing can be estimated from the breakdown and extinction voltages of the discharge. For a capacitor C charged through a resistance of R from a supply voltage V_s, the potential difference is given by

$$V(t) = V_s \left[1 - \exp\left(-\frac{t}{RC} \right) \right]. \tag{3.15}$$

Solving for t then gives

$$t(V) = RC \ln \left(\frac{V_s}{V_s - V} \right). \tag{3.16}$$

An estimation for the time period is then given by the difference between the times at which V reaches V_{br} and V_{ex}

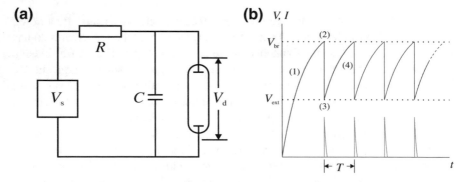

(a)   **(b)**

Fig. 3.9 a Circuit diagram of a Pearson-Anson Oscillator. **b** Voltage oscillations across the capacitor (top) and the subsequent current pulses through the discharge (bottom)

$$T = RC \ln \left(\frac{V_s}{V_s - V_{br}} \right) - RC \ln \left(\frac{V_s}{V_s - V_{ex}} \right), \tag{3.17}$$

which by the laws of logarithms can be written

$$T = RC \ln \left(\frac{V_s - V_{ex}}{V_s - V_{br}} \right). \tag{3.18}$$

If however, V_s is close to V_{br}, T will be unstable as the capacitor voltage will only reach V_{br} in its exponent tail and so will be sensitive to any minor variation in V_{br} brought about by fluctuations in temperature, pressure or some other factor.

There is an extra condition required for this flashing to occur, which is that the ballast resistance is high enough so that when the discharge is struck the current flowing through R is insufficient to maintain the discharge without the current from the capacitor. This critical resistance is given by [12]

$$R_c = \frac{V_s - V_{ex}}{\kappa(V_{ex} - V_{Cath})}, \tag{3.19}$$

where V_{Cath} is the cathode fall potential, and κ is the conductance of the discharge tube, which is highly dependent on the active area of the cathode. Typical values of R_c range from $\sim 100 \, K\Omega$ to $\sim 1 \, M\Omega$.

During the current pulse the discharge exhibits many of the characteristics of a continuous discharge, with the high currents possibly resulting in arc-like behaviour. However one of the main features of pulsed discharges are the high levels of light emitted once the discharge has been extinguished, otherwise known as the *afterglow*. The driving force behind the afterglow emission is the electron-ion recombination that occurs in the wake of the discharge.

References

1. C. S. Willett, *Introduction to Gas Lasers: Population Inversion Mechanisms*, Pergamon Press (1974)
2. A. von Engel, *Ionized Gases*, Oxford University Press (1965)
3. J.M. Somerville, A magnetic gas discharge tube oscillator. J. Sci. Instrum. **31**, 279 (1954)
4. A.W. Hull, The Dynatron: a vacuum tube possessing negative electric resistance. Proc. Inst. Radio Eng. **6**(1), 5 (1918)
5. B.J. Udelson, J.E. Creedon, Comparison of electron density in different regions of a dc glow discharge. Phys. Rev. **88**(1), 145 (1952)
6. A. von Engel, *Electric Plasmas: Their Nature and Uses*, Taylor & Francis (1983)
7. R.T. Young, Calculation of average electron energies in HeNe discharges. J. Appl. Phys. **36**(7), 2324 (1965)
8. D. Schuöcker, H. Lagger, W. Reif, Theoretical description of discharge plasma and calculation of maximum output intensity of He-Ne waveguide lasers as a function of discharge tube radius. IEEE J. Quantum Electr. **15**(4), 232 (1979)
9. P.W. Smith, A waveguide gas laser. Appl. Phys. Lett. **19**, 132 (1971)

10. P.W. Smith, P.J. Maloney, A self-stabilized 3.5-μm waveuide He-Xe laser. Appl. Phys. Lett. **22**, 667 (1973)
11. S.O. Pearson, H.S.G. Anson, The neon tube as a means of producing intermittent currents. Proc. Phys. Soc. Lond. **34**(1), 204 (1921)
12. J. Taylor, W. Clarkson, A critical resistance for flashing of the low voltage neon discharge tube. Proc. Phys. Soc. Lond. **36**(1), 269 (1923)

Chapter 4
Electrically Pumped Noble Gas Lasers

Laser systems that use noble gases as a laser medium have attracted lots of attention due to the abundance of high gain emission lines [1] and the ease of performing stable electrical discharges.

This chapter discusses the specific processes of excitation within noble gas discharge lasers and their benefits, before introducing a number of examples.

4.1 Excitation Processes

Section 3.4.1 introduced the concept of inelastic collisions and their role in the production and maintenance of a glow discharge. However while it is mainly electron ionisation collisions that drive the discharge, there are many other types of collision occurring that produce the excitations required for lasing. Those important to noble gas lasing are highlighted below.

4.1.1 Electron Impact

As electron impact collisions are the most abundant, it is of no surprise that they are the primary source of excitation as well as ionisation in a gaseous discharge. Figure 4.1 demonstrates why the noble gases are so attractive not only as the 'active' gas in a laser but also as a 'buffer' gas in metal-vapour lasers. With a high electron collision cross-section, particularly for argon, krypton and xenon, the chance of an excitation or ionisation from an electron collision is high.

A schematic energy level diagram for all the noble gases (except helium) can be seen in Fig. 4.2. The generalised laser cycle as reported by Bennett [3] is marked by

© Springer International Publishing AG, part of Springer Nature 2018
A. Love, *Hollow Core Optical Fibre Based Gas Discharge Laser Systems*,
Springer Theses, https://doi.org/10.1007/978-3-319-93970-4_4

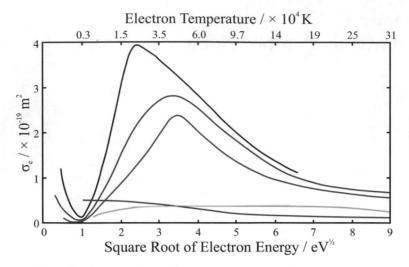

Fig. 4.1 Cross-sections for helium (red), neon (cyan), argon (blue), krypton (magenta) and xenon (black) (Recreated from [2])

a series of arrows. The black dashed arrows represent collision transitions, the solid grey lasing transitions and the dotted grey fast radiative transitions.

The general form for the excitation of neutral noble gases from the ground state to the upper lasing levels by electron impact are given by

$$(np)^6 + e^- + K.E. \; - \!\!\!\!\begin{array}{l} \nearrow (np)^5 \, (m+2)s + e^- \\ \searrow (np)^5 \, md + e^-, \end{array} \tag{4.1}$$

where $m = n, n + 1, \ldots$ and $n = 2, 3, 4$ and 5 for neon, argon, krypton and xenon respectively are the principal quantum numbers (n) and $s = 0$, $p = 1$ and $d = 2$ are the orbital quantum numbers (l). These transitions have large cross-sections as they obey the transition selection rules, namely $\Delta l = \pm 1$ and $\Delta J = 0, \pm 1$ where J is the total angular momentum [4].

With certain pressure and discharge tube diameter conditions the upper lasing state can be trapped, meaning any spontaneous emission from the excited state to the ground state is re-absorbed by another atom so the upper state remains populated. If the conditions are also suited to prevent excitation to other electronic configurations by atom-atom collisions, radiative decay can only occur through the laser transitions to the $(np)^5 \, (m+1)p$ lower lasing levels and a population inversion can be set up.

In order for this inversion to be maintained, the lower lasing levels must also be depleted quickly. As the selection rules forbid a direct transition to the ground state, this happens via a fast optical transition to the $(np)^5 \, (n+1)s$ level. With atoms in this state also trapped it is metastable, and so fast depopulation requires collisional transitions to the ground state. A laser system such as this is capable of continuous oscillation.

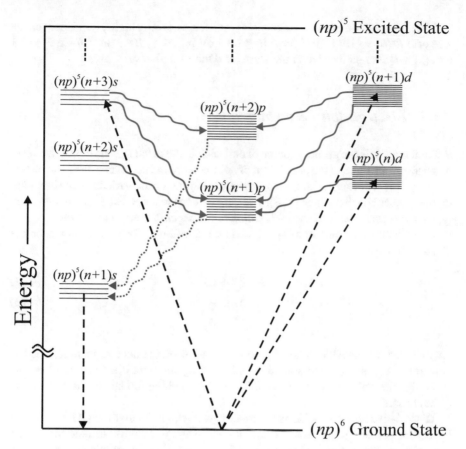

Fig. 4.2 General electronic configuration of neon, argon, krypton and xenon. The dashed lines represent collisional transitions, the dotted wavy lines are fast radiative transitions and the solid wavy lines are transitions available to lase. Note that for neon there are no $(2p)^5\ 2d$ states. (Recreated from [3])

4.1.2 Atom-Atom Excitation Transfer

A key source of excitation to the upper lasing levels in helium-neon laser systems is the resonant excitation-energy transfer that occurs between excited helium atoms and ground state neon atoms. The general form for such processes is given by

$$A^* + B \rightleftharpoons A + B^* \pm \Delta E,$$

where ΔE is the difference in potential energy between excited atomic species A^* and B^*. When A^* is metastable and B^* can decay radiatively the reaction primarily occurs in the forward direction.

The cross-section of such an excitation transfer is not only highly dependent on the size of ΔE ($\sigma_a \sim 10^{-18}$ m^{-2} for $\Delta E \leq 0.01$ eV but $\sigma_a \lesssim 10^{-26}$ m^{-2} for $\Delta E \sim 0.1$ eV) [5], but also on the size of the atoms and their relative velocity.

4.1.3   Electron-Ion Recombination

Recombination is the primary source of emission in the discharge afterglow, however it can also contribute to the population of neutral atom excited states during an active discharge [6]. The mechanisms involved in recombination are varied and the literature on the subject is rather vague [7, 8]. The general consensus is that there are two main types; *collisional radiative recombination* and *dissociative recombination*.

Collisional radiative recombination has three different forms, depending on the electron density;

$$A^+ + e^- \quad\quad \rightarrow \quad A^* + h\nu \tag{4.2a}$$

$$A^+ + e^- + e^- \rightarrow \quad A^* + e^- \tag{4.2b}$$

$$A^+ + e^- + A \quad \rightarrow \quad A^* + A. \tag{4.2c}$$

At low electron densities radiative recombination (4.2a) occurs, in high electron densities a collisional recombination stabilised by an extra electron (4.2b) occurs or in high neutral atom densities a collisional recombination stabilised by an atom (4.2c) occurs.

Dissociative recombination is the process where a molecular ion $(AB)^+$, created by earlier collisions between excited atoms, recombines with an electron before dissociating into two neutral atoms A and B, either (or both) of which can be in an excited state

$$(AB)^+ + e^- \rightarrow (AB)^* \rightarrow A^{(*)} + B^{(*)}. \tag{4.3}$$

After capturing a slow electron, the two atoms in the intermediate neutral molecular state $(AB)^*$ repel one another, becoming permanently dissociated and gaining kinetic energy. The process is represented graphically in Fig. 4.3 where an electron of energy ϵ recombines with molecular ion $(AB)^+$, existing in one of a number of states (red lines) along its potential energy curve (black line), to form unstable neutral molecule $(AB)^*$ with repulsive potential (blue line).

The major difference between the collisional electron-atomic ion recombination process, and the dissociative electron-molecular ion recombination process is that while the former can result in atoms that are in any of the available excited states, dissociative recombination can only produce atoms in excited levels below the minimum of the $(AB)^+$ ion potential curve. In other words, the dissociation will convert a minimum amount of the ionic potential energy into repulsive kinetic energy, meaning any states that have an difference in energy to the ionised state greater than this dissociation energy cannot be populated.

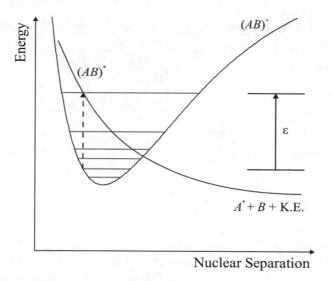

Fig. 4.3 Hypothetical potential energy curves involved in the dissociative-recombination process. (Recreated from [7])

4.2 Gain and the *pD* Product

A notable characteristic of neutral noble gas laser systems is the inverse relationship between the gain and the laser tube diameter for a constant pD value (and hence constant electron temperature) [9–11]. A number of different explanations for this behaviour have been put forward through the years.

Bennett [3] first proposed that as the lower $(np)^5(n+1)s$ states can only be de-excited via diffusion to the tube walls, reducing the tube diameter to increase the number of wall collisions will boost the depopulation of the lowest state. However, Willett [7] argues that as the number of excited atoms lost per second per unit length of plasma to the tube walls Q can be given by

$$Q \propto \frac{\Delta N}{pD}, \tag{4.4}$$

where ΔN is the difference in concentration of atoms between the centre and on the inside wall of the discharge tube, then keeping pD constant despite varying D has no effect on the depopulation rate. Instead Willett proposed that this increase in gain is purely a result of the increased pressure required to maintain a constant pD value, which will increase the population of all the energy levels involved [7].

Table 4.1 Mid-IR xenon laser lines of interest and their associated energy levels in Racah notation

Wavelength/μm	Transition
2.026	$5d\left[\frac{3}{2}\right]_1^o \to 6p\left[\frac{3}{2}\right]_1$
3.107	$5d\left[\frac{5}{2}\right]_3^o \to 6p\left[\frac{3}{2}\right]_2$
3.367	$5d\left[\frac{5}{2}\right]_2^o \to 6p\left[\frac{3}{2}\right]_1$
3.508	$5d\left[\frac{7}{2}\right]_3^o \to 6p\left[\frac{5}{2}\right]_2$
5.575	$5d\left[\frac{7}{2}\right]_4^o \to 6p\left[\frac{5}{2}\right]_3$

4.3 The Xenon Laser

With a high electron collision cross-section, low ionisation energy and an abundance of high gain emission lines [12], xenon gas is the most logical starting point for any new experimental laser system. The first xenon based lasers operated on the 2.026 μm line, although since then many more lines capable of demonstrating lasing have been discovered throughout the mid-IR spectral region with the 5.575 μm line exhibiting the highest gain. However it is the high gain lines that fall between 3 and 4 μm where the newly developed hollow core optical fibres (see Chap. 5) can provide excellent guidance that are of particular interest to this project.

All the high gain lines listed in Table 4.1 involve transitions between the upper $(5p)^5\,5d$ and the lower $(5p)^5\,6p$ energy levels, which can be seen in Fig. 4.4. The energy levels involved are given in Racah notation, defined in the front matter. The 3.11, 3.37 and 3.51 μm lines have been highlighted as they exhibit particularly high gain, with the latter having been shown to produce optical gains of up to 60 dBm^{-1}.

Contrary to the generalised system pictured by Bennett discussed in Sect. 4.1.1, the only upper lasing level connected to the ground state via an optically allowed transition is the $5d\left[\frac{3}{2}\right]_1^o$ level of the 2.026 μm line. The other four lines listed in Table 4.1 all feature a total angular momentum of $J \geq 2$, and so are forbidden optically (in the ground state, $J = 0$). This does not mean that they cannot be directly excited by electron impact, but collisional excitations of this type will suffer from lower cross-sections.

These states are instead mainly populated by a cascade of collisional transitions from the higher states which are more favourably directly excited, with transitions to the $5d\left[\frac{7}{2}\right]_3^o$ state of the 3.508 μm line featuring a higher statistical weight [13]. Given favourable discharge conditions (pressure, tube diameter etc.) these states can build up a high population, and with no resonant emission to the ground state possible the lasing process proceeds largely as described earlier in Sect. 4.1.1.

In addition to electron impact excitation, the upper levels are also populated through recombination processes. As the ionisation energy of xenon and dissociation energy for the ground state of Xe_2^+ are equal to 12.1 and 1 eV respectively [8], any states below 11.1 eV can be populated by dissociative recombination. This includes all the $5d$ upper laser levels which sit between 10.4 and 9.5 eV, so both electron-atomic

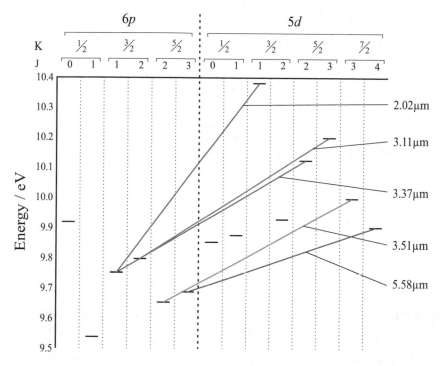

Fig. 4.4 Partial energy level diagram for xenon featuring a selection of the major mid-IR laser transition lines

ion and electron-molecular ion recombination can be responsible for populating these states. As with electron impact excitation, electron-ion recombination does not necessarily directly excite the favoured upper state but rather the states above, resulting in a similar cascade of collisional transitions.

While xenon exhibits high gain on its own, the addition of helium can increase said gain even further. This is not a result of atom-atom excitation transfer like in helium-neon lasers (see Sect. 4.4) as the metastable $^2S_{\frac{1}{2}} \, 2s[\frac{1}{2}]_1$ helium state sits at a much higher energy than the $5d$ xenon states (as shown in Fig. 4.5). Instead it is a combination of many factors.

In earlier works the helium was purely considered to act as a stabiliser, reducing the electron temperature while increasing the electron population [7]. However it has since been shown to play a more active role in both producing more xenon ions which can then recombine [6], and in the quenching of the lower laser levels [14].

It is however uncommon to use a helium-xenon mixture in CW DC-excited system due to the strong cataphoretic effects that can occur; a rapid clean up of the gases is often essential. Alternatively, a pulsed DC source can be used.

Due to the requirement of low pressure to easily achieve breakdown, early xenon lasers that were pumped by longitudinal gas discharges were limited in power by the small number of atoms involved. In a 250 : 1 helium-xenon laser at 5 torr (6.7 mbar),

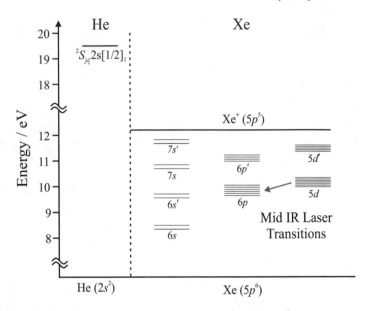

Fig. 4.5 Partial energy level diagram for helium-xenon. As the $^2S_{\frac{1}{2}}\,2s[\frac{1}{2}]_1$ helium state lies ≈ 9 eV higher than the $5d$ xenon states, the collisional cross section for an atom-atom excitation transfer is very low. The $^2S_{\frac{1}{2}}\,2s[\frac{1}{2}]_1$ helium state is however higher than the xenon first ionisation energy of 12.1 eV, meaning collisions between the two atoms can result in a Penning ionisation collision

the achievable CW power on the 2.026 μm line was ≈ 10 mW in a 2.25 m long, 8 mm wide discharge tube [3].

Higher power xenon lasers have been achieved by using much higher pressures, these however required pumping transversely to keep the breakdown potentials achievable. Pulsed discharges of 100 ns in duration within a 2 m long, 2.5 cm wide laser tube achieved peak powers of up to 1 kW in a 20:1 helium-xenon mixture at 250 torr (330 mbar) [15]. Higher powers have subsequently been measured in high pressure Ar-Xe mixtures, pumped by microwave excitation [16] or an electron gun [17–19]. The implementation of such mechanisms into a hollow core fibre design is however a far greater engineering challenge than the methods proposed in this thesis.

4.4 The Helium-Neon Laser

Unlike xenon, the addition of helium to neon is almost essential to the operation of a laser on the neon lines. This is due to the very strong atom-atom excitation transfer between the metastable $^2S_{\frac{1}{2}}\,2s[\frac{1}{2}]_1$ and $^2S_{\frac{1}{2}}\,2s[\frac{1}{2}]_0$ helium levels and the upper lasing levels of neon (see Fig. 4.6).

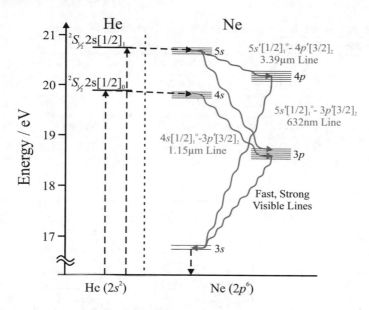

Fig. 4.6 Partial energy level diagram for helium-neon. The coincident energies of the $^2S_{\frac{1}{2}}$ $2s[\frac{1}{2}]_1$ and $^2S_{\frac{1}{2}}$ $2s[\frac{1}{2}]_0$ helium and the $5s$ and $4s$ neon levels respectively provide high collisional cross sections for atom-atom excitation transfer, allowing for fast population of these upper states and lasing on a number of lines

The very first gas laser system was built upon helium and neon as a gain medium, and demonstrated lasing on five lines around 1.15 μm [20]. He-Ne systems are very popular due to the strong visible 632.8 nm line. However to achieve lasing on this line the very high gain 3.39 μm line with which it shares an upper energy level has to be suppressed.

4.5   Neutral and Ionised Argon Lasers

There are four principle neutral argon laser lines in the infrared listed below in Table 4.2, all of which occur between the $(3p)^5$ $3d$ and $(3p)^5$ $4p$ levels. As with xenon, the addition of helium to the mix can lead to an increase of gain.

Table 4.2 Mid-IR neutral argon laser lines of interest and their associated energy levels in Racah notation

Wavelength/μm	Transition
1.792	$3d\left[\frac{1}{2}\right]_1^o \rightarrow 4p\left[\frac{1}{2}\right]_1$
2.062	$3d\left[\frac{3}{2}\right]_2^o \rightarrow 4p'\left[\frac{3}{2}\right]_2$
2.208	$3d\left[\frac{1}{2}\right]_1^o \rightarrow 4p'\left[\frac{3}{2}\right]_2$
2.397	$3d\left[\frac{1}{2}\right]_0^o \rightarrow 4p'\left[\frac{1}{2}\right]_1$

Table 4.3 Ar-II laser lines and their associated energy levels in Racah notation

Wavelength/nm	Transition	Wavelength/nm	Transition
454.5	$^3P_2\, 4p\, [1]^o_{\frac{3}{2}} \rightarrow {}^3P_2\, 4s\, [2]_{\frac{3}{2}}$	488.0	$^3P_2\, 4p\, [3]^o_{\frac{5}{2}} \rightarrow {}^3P_2\, 4s\, [2]_{\frac{3}{2}}$
457.9	$^3P_1\, 4p\, [1]^o_{\frac{1}{2}} \rightarrow {}^3P_1\, 4s\, [1]_{\frac{1}{2}}$	496.5	$^3P_1\, 4p\, [1]^o_{\frac{3}{2}} \rightarrow {}^3P_1\, 4s\, [1]_{\frac{1}{2}}$
465.8	$^3P_2\, 4p\, [1]^o_{\frac{3}{2}} \rightarrow {}^3P_2\, 4s\, [2]_{\frac{3}{2}}$	501.7	$^1D_2\, 4p\, [3]^o_{\frac{5}{2}} \rightarrow {}^3P_2\, 3d\, [1]_{\frac{3}{2}}$
472.7	$^3P_1\, 4p\, [1]^o_{\frac{3}{2}} \rightarrow {}^3P_2\, 4s\, [2]_{\frac{3}{2}}$	514.5	$^3P_1\, 4p\, [2]^o_{\frac{5}{2}} \rightarrow {}^3P_2\, 4s\, [2]_{\frac{3}{2}}$
476.5	$^3P_2\, 4p\, [1]^o_{\frac{3}{2}} \rightarrow {}^3P_1\, 4s\, [1]_{\frac{1}{2}}$	528.7	$^3P_0\, 4p\, [1]^o_{\frac{3}{2}} \rightarrow {}^3P_1\, 4s\, [1]_{\frac{1}{2}}$

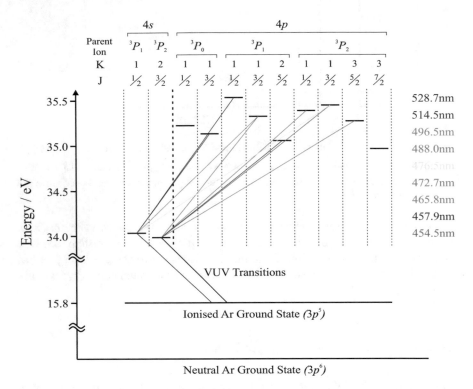

Fig. 4.7 Partial energy level diagram for singly ionised argon featuring nine of the ten visible laser transitions

In addition to these neutral laser lines, laser oscillation is also possible on nine of the visible lines between the $4p$ and $4s$ energy levels and one of the $4p$ to $3d$ lines (listed in Table 4.3) in singly ionised argon. Excitation to the upper lasing level occurs via electron impact in either a single step process, where a single collisional event excites a ground state neutral argon atom, or two step process where a neutral atom is first ionised, then excited from its ionic ground state to the upper level in two separate collisions.

As the energies of the ionised upper levels are quite high compared to the neutral upper levels (around 35 eV compared to 10 eV as shown in Fig. 4.7), ion lasers generally operate at higher current densities than their neutral counterparts with typical values of at least $\approx$500 A cm^{-2} pulsed and $\approx$100 A cm^{-2} CW [21].

References

1. C.S. Willett, Laser lines in atomic species. Prog. Quantum Electron. **1**, 273 (1971)
2. R.B. Brode, The quantitative study of the collisions of electrons with atoms. Rev. Mod. Phys. **5**(4), 257 (1933)
3. W.R. Bennett Jr., Gaseous optical masers. Appl. Opt. **1**(S1), 24 (1962)
4. W. Demtröder, *Atoms, Molecules and Photons: An Introduction to Atomic-, Molecular-, and Quantum-Physics* (Springer, 2010)
5. N.F. Mott, H.S.W. Massey, *The Theory of Atomic Collisions* (Oxford University Press, 1952)
6. R. Shuker, A. Szöke, E. Zamir, Y. Binur, Energy transfer in noble-gas mixtures: penning ionization in He/Xe. Phys. Rev. A **11**(4), 1187 (1975)
7. C.S. Willett, *Introduction to Gas Lasers: Population Inversion Mechanisms* (Pergamon Press, 1974)
8. A. Barbet, N. Sadeghi, J.C. Pebay-Peyroula, Study of the electron-ion recombination processes in the xenon afterglow plasma. J. Phys. B At. Mol. Phys. **8**(10), 1785 (1975)
9. C. Patel, W. Bennett, W. Faust, R. McFarlane, Infrared spectroscopy using stimulated emission techniques. Phys. Rev. Lett. **9**(3), 102 (1962)
10. P.W. Smith, On the optimum geometry of a 6328Å laser oscillator. IEEE J. Quantum Electron. **2**(4), 77 (1966)
11. P.O. Clark, Investigation of the operating characteristics of the 3.5 μm xenon laser. IEEE J. Quantum Electron. **1**(3), 109 (1965)
12. W.L. Faust, R.A. McFarlane, C.K.N. Patel, C.G.B. Garrett, Gas maser spectroscopy in the infrared. Appl. Phys. Lett. **1**(4), 85 (1962)
13. A.V. Karelin, O.V. Simakova, Kinetics of the active medium of a multiwave IR xenon laser in hard-ioniser-pumped mixtures with He and Ar. I. Electron-beam pumping. Quantum Electron. **29**(8), 678 (1999)
14. W.J. Alford, State-to-state rate constants for quenching of xenon 6p levels by rare gases. J. Chem. Phys. **96**(6), 4330 (1992)
15. S.E. Schwarz, T.A. DeTemple, R. Targ, High-pressure pulsed xenon laser. Appl. Phys. Lett. **17**(7), 305 (1970)
16. C.L. Gordon, C.P. Christensen, B. Feldman, Microwave-discharge excitation of an ArXe laser. Opt. Lett. **13**(2), 114 (1988)
17. S.A. Lawton, J.B. Richards, L.A. Newman, L. Specht, T.A. DeTemple, The high pressure neutral infrared xenon laser. J. Appl. Phys. **50**(6), 3888 (1979)
18. L.N. Litzenberger, D.W. Trainor, M.W. McGeoch, A 650 J e-beam pumped atomic xenon laser. IEEE J. Quantum Electron. **26**(9), 1668 (1990)
19. J.P. Apruzese, J.L. Giuliani, M.F. Wolford, J.D. Sethian, G.M. Petrov, D.D. Hinshelwood, M.C. Myers, A. Dasgupta, F. Hegeler, T. Petrova, Optimizing the ArXe infrared laser on the Naval research laboratory's electra generator. J. Appl. Phys. **104**(1), 13101 (2008)
20. A. Javan, W.R. Bennett, D.R. Herriott, Population inversion and continuous optical maser oscillation in a gas discharge containing a He-Ne mixture. Phys. Rev. Lett. **6**(3), 106 (1961)
21. W.B. Bridges, A.N. Chester, J.V. Parker, A.S. Halsted, Ion laser plasmas. Proc. IEEE **59**(5), 724 (1971)

Chapter 5
Introduction to Optical Fibres

The concept of using a glass fibre to transmit light from one place to another dates back to as early as the start of the twentieth century [1]. As technologies improved throughout the century more applications for optical fibres were developed, with the uses of modern fibres stretching across a wide range of fields including lasing, sensing, imaging, computing and telecommunications.

This chapter introduces the fundamental concept of optical fibres, including their structure and the mechanism with which they guide light, before developing these into the more complex designs of hollow cored fibres.

5.1 Step-Index Fibres

Conventional optical fibres are typically *step-index fibres* (see Fig. 5.1). These are composed of a silica glass core surrounded by a silica glass cladding of a lower refractive index. This refractive index difference is achieved by doping the glass with either germanium and phosphorous to raise the index, or flourine and boron to decrease the index. The refractive index drop at the core-cladding boundary allows for *total internal reflection* (TIR) to occur, and so light can be trapped within the core for the length of the fibre.

The idea of a ray of light propagating down a fibre by continuously bouncing off of this boundary wall is however only valid if the core diameter is much greater than the wavelength of the propagating light. For a typical step-index fibre with a core size of $10\,\mu$m, a micron-diameter beam of light with a wavelength of 1550 nm (the standard telecommunications wavelength) has a Rayleigh length (defined later in Sect. 6.1.6) of just $0.5\,\mu$m. As this is much shorter than the propagation distance, the position of the ray cannot be well defined and is hence unphysical. A more complete description of how light travels down an optical fibre requires the use of waveguide modes.

5.2 Waveguide Modes

A waveguide mode is defined as an electromagnetic field that only changes in phase as it propagates through a waveguide, or in mathematical terms a field with the form

© Springer International Publishing AG, part of Springer Nature 2018

A. Love, *Hollow Core Optical Fibre Based Gas Discharge Laser Systems*,

Springer Theses, https://doi.org/10.1007/978-3-319-93970-4_5

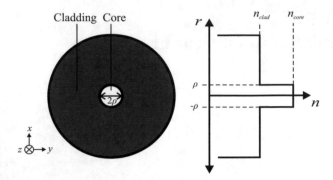

Fig. 5.1 Cross-section of a step index fibre with core radius ρ

$$E(x, y, z, t) = E(x, y)e^{i(\beta z - \omega t)}, \tag{5.1}$$

where (x, y) define the plane parallel to the cross-sectional surface of the fibre, and z is in the direction of propagation. The *propagation constant* β is then defined as the z component of the wavevector. A useful analogy to waveguide modes is the finite potential well in quantum physics [2], as shown in Fig. 5.2.

In the finite potential well problem, there are a finite number of bound states with energies less than the depth of the well which vary only in phase with time. These states are then analogous to modes within a fibre, with the negative potential well replaced with a positive index step, and the energy eigenvalues by the modal propagation constant. As each mode has its own propagation constant, the *modal index* can be defined as

$$n_{mode} = \frac{\beta}{k}, \tag{5.2}$$

where k is the free space wave vector. Continuing with the analogy, there is a finite number of modes that can be confined within the step index fibre, with values of β such that

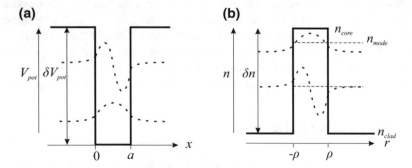

Fig. 5.2 **a** Solutions to the Schrödinger equation in a 1D finite potential well, where a is the width of the well and δV_{pot} is the potential drop, and **b** solutions to the Maxwell equations for a step-index fibre

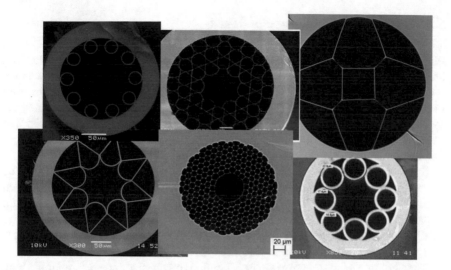

Fig. 5.3 Micrographs of a number of hollow core fibre designs. Clockwise from top left: a ten resonator free boundary fibre* [5], a kagomé cladding structure fibre* [6], a square core anti-resonant fibre [7], an early tube lattice negative curvature fibre [8], a hollow photonic bandgap fibre [9], a negative curvature anti-resonant guiding fibre* [10]. Those marked with an asterisk were fabricated at the University of Bath

$$kn_{clad} < \beta \leq kn_{core}. \tag{5.3}$$

A fibre that supports only one spacial mode is known as a *single mode fibre* (SMF) while a fibre that supports more than one spacial mode is known as a *multimode fibre*. The mode with the highest value of β is referred to as the *fundamental mode* (shown at top of Fig. 5.2b), and is the only mode supported in SMF. It is important to note that in truth a single mode fibre guides two modes, as there are two orthogonal polarisation modes described by the fundamental spacial mode (Fig. 5.3).

5.3 Hollow Core Fibres

The concept of a hollow core optical fibre (HCF) more sophisticated than a simple capillary was first demonstrated in 1999 [3]. The advantages of HCFs include the reduced infrared absorption in air over silica, making it possible to guide light at mid-infrared wavelengths, and the ability to fill the fibre with a controlled pressure of gas. The use of a gas as a guiding medium not only introduced the possibility of building a gas laser, but also gave a much greater flexibility over the nonlinear optical effects possible [4].

Contrary to the solid fibres described above, hollow core fibres feature a refractive index increase from the air core to the glass cladding, and so their guidance can no longer be explained by TIR. Instead there are two new methods of confinement; the use of a *photonic bandgap* (PBG), or of *anti-resonant reflections* (AR).

5.3.1 Photonic Bandgap Guidance

Photonic bandgap guidance uses a periodic structure of silica in air as a cladding around an air core. The periodic array of high index index regions within a low index one acts as a 2D *photonic crystal* and can forbid certain wavelengths of light from passing through it. The physics of a photon faced with a photonic crystal is analogous to that of the tight binding model for an electron in an array of atoms [11]. In that case, the periodic potential introduced by the atoms causes distortions in the electron dispersion relation, producing *bandgaps* where no electron states can exist.

In the photonic case, the periodic high index regions act in the same way as these atomic potentials, producing bandgaps (or stop-bands in the 1D case, shown in Fig. 5.4) in the photon dispersion relation forbidding certain frequencies of light to pass through.

Photonic bandgap guidance is also possible in an all solid fibre, with a periodic array of high index doped rods in pure silica as a cladding [12].

The photonic band structure is very dependent on the size and spacing of the high index regions, so the practicality of building the cladding structure limits the size of the fibre core, with some of the largest reported at 50 μm [9]. Such core sizes can present great difficulties when trying to produce sustainable gas discharges, making the larger core sizes available in AR guiding fibres more attractive for constructing a hollow fibre based gas discharge laser.

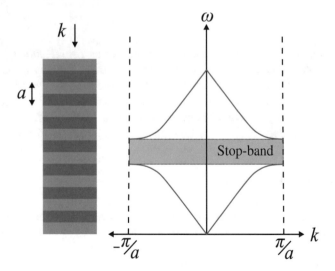

Fig. 5.4 A '1D photonic crystal' and its dispersion relation. The periodic array of high index regions (dark grey) within a low index region (light grey) create a photonic stop-band (the 1D equivalent of a bandgap) in the dispersion relation, preventing light of certain frequencies from passing through the crystal

5.3.2 *Anti-resonant Guidance*

Resonance and anti-resonance refer to the physical phenomena that occur when light encounters a thin film of dielectric material. Each surface of the film acts as a partially reflective mirror so as a beam of light passes through, a fraction of that light is reflected at each interface. With reflections at both boundaries, a secondary beam that has split from the initial beam at the rear boundary can then make an integer number of additional trips back and forth through the film, before arriving back at the rear boundary with an added phase difference of

$$\delta_{ph} = \frac{4\pi}{\lambda} n d M \cos \theta,$$
(5.4)

to the primary beam. Here n and d are the refractive index and thickness of the film respectively, θ the angle of incidence and M the number of round trips the secondary beam has taken through the film.

For wavelengths λ_{cons} where this phase difference is equal to $2N\pi$ (for any integer N), the outgoing primary and secondary beams are in phase and constructively interfere, as is shown in Fig. 5.5a. For the narrow band of *resonant* wavelengths around λ_{cons} the transmission of the film is high. However for all other wavelengths the primary and secondary beams are out of phase and destructively interfere, resulting in strong *anti-resonant* reflections from the film. This effect is commonly observed as brightly coloured patterns visible in soap bubbles.

The use of these anti-resonant reflections in waveguiding was first demonstrated in 1986 in SiO_2-Si planar waveguides [13], where light was guided in a layer of pure SiO_2 by TIR at the air$-SiO_2$ boundary above and anti-resonant reflections with a thin polycrystalline Si layer below. The transmission of these waveguides featured narrow high loss resonant wavelengths surrounded by broadband low loss anti-resonant wavelength regions.

The position of these resonant bands can be estimated by the anti-resonant reflecting optical waveguide (ARROW) model

$$\lambda_{res} = \frac{2n_{core}d}{m} \left[\left(\frac{n_{wall}}{n_{core}} \right)^2 - 1 \right]^{\frac{1}{2}},$$
(5.5)

where d is the anti-resonant layer thickness, m an integer and $n_{core,wall}$ the refractive indices for the waveguide core and anti-resonant layer respectively [14]. An example of the transmission for a waveguide with 3 μm thick walls can been seen in Fig. 5.6.

The use of the anti-resonant guiding mechanism in hollow core optical fibres stems from many different designs. To aid in the fabrication of hollow core bandgap fibres, the core-cladding boundary was often terminated with a thin silica tube (bottom middle in Fig. 5.3). The thickness of this ring was observed to have a significant impact on the fibre's guidance, with a stronger confinement occurring when the anti-resonant wavelengths of the ring fell into the guided band of the fibre [15].

(a)   **(b)**

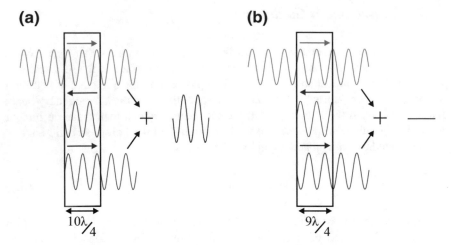

Fig. 5.5 A representation of thin film interference for a beam at normal incidence. A partial reflection (middle) from a primary beam (top) at the rear boundary is reflected back to exit the film (bottom) with a phase difference $\delta_{ph} = \frac{4\pi}{\lambda} nd$. In **a** the beams are in phase, constructive interference occurs and there is high transmission. In **b** the beams are exactly out of phase, destructive interference occurs and there is a strong reflection

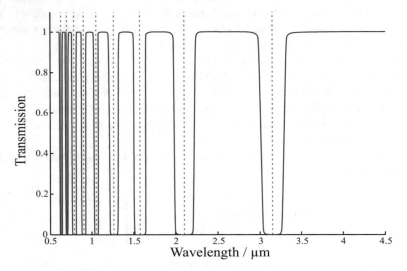

Fig. 5.6 Resonant wavelengths (dashed lines) calculated from Eq. 5.5 of the ARROW model for a hollow waveguide with 3 μm thick silica walls, and an schematic example of the ideal resulting waveguide transmission (solid line)

A second precursor to the contemporary anti-resonant designs was the 'kagomé' structure, first reported in 2002 [16]. Delivering a broader spectral range than bandgap fibres, the kagomé fibres were the first to demonstrate light guidance in air without supporting photonic bandgaps [17].

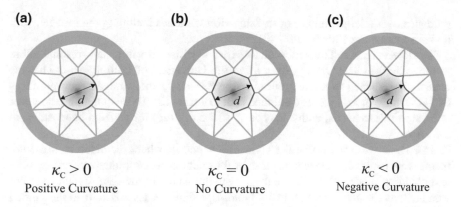

Fig. 5.7 Comparison of three anti-resonant guiding hollow core fibre structures with equal core diameters, but with **a** $\kappa_C > 0$, **b** $\kappa_C = 0$ and **c** $\kappa_C < 0$, where the inner wall curvature κ_C is defined as the derivative of the unit tangent vector normalised to the arc length. The guided mode is shown in the centre, and as κ_C decreases the mode overlap with the inner wall also decreases

It was later shown that the guidance mechanisms of kagomé fibres were almost entirely dependent on the anti-resonant reflections of the core surround, not the cladding [18]. Subsequently in 2010 a fibre that guides light purely by anti-resonant reflections with a minimum loss of 200 dBkm^{-1} within a 200 nm transmission window in the visible was demonstrated [19].

While the reliance on only a single thin wall around the fibre core introduces more simplicity into the fibre design, it is not without its drawbacks. On top of the fundamental surface scattering loss that affects all hollow core fibre designs, anti-resonant guiding fibres also suffer from a leakage loss due to the imperfect nature of the anti-resonant reflections. This means that the so called 'leaky' guided modes are strictly speaking not modes at all, but instead take the form of

$$E(x, y, x, t) = E(x, y)e^{i(\beta z - \omega t)}e^{-qz}, \tag{5.6}$$

where q is a loss constant.

These leaky modes have been shown to have as much as a 0.5% overlap with the cladding structure. This leads to an increased leakage loss due to coupling with the modes of the cladding structure as well as significant material absorption losses in the long wavelength mid-IR spectral region.

One way to reduce this modal overlap is to change the shape of the core boundary. Figure 5.7 introduces the concept of *negative curvature*, an effective way of reducing this overlap. Through optimum design of the core wall shape and core diameter, this overlap can be reduced to as low as $< 0.01\%$ [20].

The first fibres to feature this negative curvature where of the kagomé type (top middle in Fig. 5.3), and exhibited a loss of 180 dB km^{-1} in a guidance window from 0.85 μm to 1.7 μm [6]. The design was simplified in 2011 by Pryamikov et al. to an octagonal array of touching capillaries (bottom right in 5.3), which exhibited guidance over a large number of transmission windows including the mid-IR, with losses approaching 100 dB km^{-1} [8].

The lowest recorded transmission loss for a negative curvature, anti-resonant fibre in the mid-IR to date was by Yu et al. at the University of Bath in 2012. The 'ice cream cone' shaped structure (bottom left in Fig. 5.3) demonstrated a minimum loss of 34 dB km^{-1} at 3.04 μm and guidance out to 4 μm [10]. These fibres were however quite sensitive to bends, with a loss of up to 5.5 dB/turn for an 8 cm bend radius at 3.35 μm.

In 2013 it was shown that the connecting 'nodes' where the cladding structure meets were increasing the leakage loss of these fibres via coupling between the core and cladding modes [21]. This leakage was dramatically increased when the fibre was bent, as the act of mechanically bending a fibre causes a β shift in the guided mode of the node sites, increasing the coupling to the core and subsequently the fibre loss.

By disconnecting the capillary cladding structure of Pryamikov and creating a 'free' core boundary this effect could be eliminated. One such example of a *free boundary fibre* developed at the University of Bath by Belardi et al. has demonstrated a minimal loss of 100 dB km^{-1} with a bend loss of 0.2 dB/turn for an 8 cm bend radius at 3.35 μm [5].

These new free boundary fibres possessed core sizes of over 100 μm, which on top of the low transmission and bend losses they were the perfect candidate for the development of a new generation of gas lasers.

5.4 Optical Fibre Fabrication

5.4.1 Fabrication Basics

The fabrication of a standard optical fibre starts with a glass *preform*; a rod of glass usually around 20 to 30 mm in diameter and 1 to 2 m in length with the desired cross-sectional dopant structure. This preform is held vertically at the top of a *fibre drawing tower*, shown schematically in Fig. 5.8, where it is lowered into a furnace and heated to around 2000°C.

At these temperatures the glass becomes soft, allowing it to be pulled out of the bottom of the furnace. By the laws of mass conservation, if the glass preform is pulled out of the furnace faster than it is fed into it, its diameter is reduced while the cross-sectional structure is maintained. This process of pulling a preform down to size is referred to as *drawing*. The industrial standard size of fibre used for telecommunication is 125 μm in diameter, although this can vary depending on the use.

While remarkably strong, these fibres are very susceptible to breakages from abrasions and contaminants, so for extra protection they are covered in a UV curable polymer as they come out of the furnace before being wound onto a drum.

The whole process is semi-automated, with a laser diameter monitor for the fibre and precisely controlled draw and feed speeds to achieve the correct diameter.

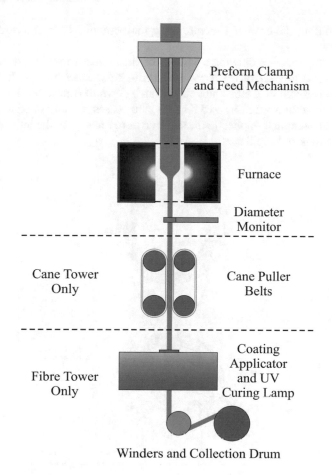

Fig. 5.8 Schematic diagram of a fibre drawing tower. The top section of a clamp, feed mechanism, furnace and diameter monitor is universal in all applications, however the pulling mechanism varies depending on the desired diameter after the draw. For larger scale draws (to ∼1 mm) such as when drawing rods, capillaries or canes (see Sect. 5.4.2), the middle section which utilises a belt system is used. For drawing fibre the lower section is used to coat the fibres and wind them onto a drum

5.4.2 Fabricating Microstructured Fibres: The 'Stack and Draw' Method

The fibres pictured in Fig. 5.3 are some examples of so-called microstructured fibres, a term that refers to any optical fibre that features some sort of cross-sectional design beyond that of a step- or graded-index fibre. Other examples include photonic crystal fibres (PCFs) [22] and multicore fibres [23, 24].

The most common practice of creating these micron-scale structures is the *stack and draw* method (illustrated in Fig. 5.9 for a free boundary fibre), a process that involves building the desired structure on a macroscopic scale before drawing it down to size.

The first step involves drawing a number of silica tubes (and rods, depending on the fibre design) to sizes of the order ~ 1 mm to be *stacked* into the fibre design. The stack is jacketed by a silica tube then drawn down to *canes*, the sizes of which can vary from around 1 mm up to 5 mm. In some cases, usually in solid-core fibre stacks, a light vacuum is applied to the stack so as to remove all the interstitial holes between the rods and capillaries.

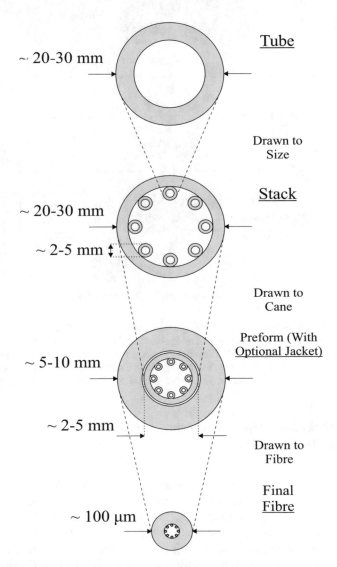

Fig. 5.9 Schematic representation of the stack and draw process

The final step is to draw these canes down to fibre, but before then the canes can be jacketed by another silica tube so that the microstructure can be drawn to the correct size while producing a fibre with a sensible outer diameter. During this final draw, pressure may be applied to any air holes within the fibre to keep them from collapsing and/or inflate them, and a vacuum is applied outside again to remove any gaps between the cane and jacket. As the fibre is drawn it is coated in a polymer as with the step index fibre.

5.4.3 Differences in Fabricating AR Fibre

The stacking process involved in the fabrication of the anti-resonant guiding fibres shown in Fig. 5.3 is slightly move involved than implied by Fig. 5.9. For a start, it is less of a 'stacking' process and more of a 'threading' process, with the capillaries being threaded into the jacket tube one by one instead of stacked on a rig.

The stack is also not the same throughout the length of jacket; the structure required in the fibre is not self-supporting and so needs some packer rods to stay together. These packers however are extremely undesirable in the fibre and so only feature in the first 20–25 cm of the ends of the stack as in Fig. 5.10. This unfortunately means that at least a third of the stack is scrap. The rest of the process is essentially the same, with the stack being drawn down to canes, to be then drawn to fibre with an optional jacketing tube for support.

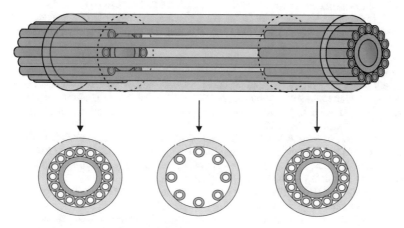

Fig. 5.10 Schematic diagram of the stack for a typical AR guiding fibre

References

1. J. Hecht, *City of Light*, Oxford University Press (1999)
2. A. I. M. Rae, *Quantum Mechanics*, Taylor & Francis (2008)
3. R.F. Cregan, Single-mode photonic band gap guidance of light in air. Science **285**(5433), 1537 (1999)
4. P.S.J. Russell, P. Hölzer, W. Chang, A. Abdolvand, J.C. Travers, Hollow-core photonic crystal fibres for gas-based nonlinear optics. Nat. Photonics **8**(4), 278 (2014)
5. W. Belardi, J.C. Knight, Hollow antiresonant fibers with low bending loss. Opt. Express **22**(8), 10091 (2014)
6. Y.Y. Wang, N.V. Wheeler, F. Couny, P.J. Roberts, F. Benabid, Low loss broadband transmission in hypocycloid-core Kagome hollow-core photonic crystal fiber. Opt. Lett. **36**(5), 669 (2011)
7. A. Hartung, J. Kobelke, A. Schwuchow, K. Wondraczek, J. Bierlich, J. Popp, T. Frosch, M.A. Schmidt, Double antiresonant hollow core fiber guidance in the deep ultraviolet by modified tunneling leaky modes. Opt. Express **22**(16), 19131 (2014)
8. A.D. Pryamikov, A.S. Biriukov, A.F. Kosolapov, V.G. Plotnichenko, S.L. Semjonov, E.M. Dianov, Demonstration of a waveguide regime for a silica hollow-core microstructured optical fiber with a negative curvature of the core boundary in the spectral region > 3.5 μm. Opt. Express **19**(2), 1441 (2011)
9. N.V. Wheeler, A.M. Heidt, N.K. Baddela, E.N. Fokoua, J.R. Hayes, S.R. Sandoghchi, F. Poletti, M.N. Petrovich, D.J. Richardson, Low-loss and low-bend-sensitivity mid-infrared guidance in a hollow-core-photonic-bandgap fiber. Opt. Letters **39**(2), 295 (2014)
10. F. Yu, W.J. Wadsworth, J.C. Knight, Low loss silica hollow core fibers for 3–4μm spectral region. Opt. Express **20**(10), 11153 (2012)
11. J. R. Hook, H. E. Hall, *Solid State Physics*, Wiley (1995)
12. F. Luan, A.K. George, T.D. Hedley, G.J. Pearce, D.M. Bird, J.C. Knight, P.S.J. Russell, All-solid photonic bandgap fiber. Opt. Lett. **29**(20), 2369 (2004)
13. M.A. Duguay, Y. Kokubun, T.L. Koch, L. Pfeiffer, Antiresonant reflecting optical waveguides in SiO2-Si multilayer structures. Appl. Phys. Lett. **49**(1), 13 (1986)
14. N.M. Litchinitser, S.C. Dunn, B. Usner, B.J. Eggleton, T.P. White, R.C. McPhedran, C.M. de Sterke, Resonances in microstructured optical waveguides. Opt. Express **11**(10), 1243 (2003)
15. P.J. Roberts, D.P. Williams, B.J. Mangan, H. Sabert, F. Couny, W.J. Wadsworth, T.A. Birks, J.C. Knight, P.S. Russell, Realizing low loss air core photonic crystal fibers by exploiting an antiresonant core surround. Opt. Express **13**(20), 8277 (2005)
16. F. Benabid, J.C. Knight, G. Antonopoulos, P.S.J. Russell, Stimulated Raman scattering in hydrogen-filled hollow-core photonic crystal fiber. Science **298**(5592), 2000 (2002)
17. F. Benabid, P.J. Roberts, Linear and nonlinear optical properties of hollow core photonic crystal fiber. J. Mod. Opt. **58**(2), 87 (2011)
18. S. Février, B. Beaudou, P. Viale, Understanding origin of loss in large pitch hollow-core photonic crystal fibers and their design simplification. Opt. Express **18**(5), 5142 (2010)
19. F. Gérôme, R. Jamier, J.-L. Auguste, G. Humbert, J.-M. Blondy, Simplified hollow-core photonic crystal fiber. Opt. Lett. **35**(8), 1157 (2010)
20. W. Belardi, J.C. Knight, Effect of core boundary curvature on the confinement losses of hollow antiresonant fibers. Opt. Express **21**(19), 21912 (2013)
21. A.N. Kolyadin, A.F. Kosolapov, A.D. Pryamikov, A.S. Biriukov, V.G. Plotnichenko, E.M. Dianov, Light transmission in negative curvature hollow core fiber in extremely high material loss region. Opt. Express **21**(8), 9514 (2013)
22. J.C. Knight, Photonic crystal fibres. Nature **424**(6950), 847 (2003)
23. K. Saitoh, S. Matsuo, Multicore fiber technology. J. Lightwave Technol. **34**(1), 55 (2016)
24. J.M. Stone, H.A.C. Wood, K. Harrington, T.A. Birks, Low index contrast imaging fibers. Opt. Lett. **42**(8), 1484 (2017)

Chapter 6
Experiment Assembly and CW Measurements of the He-Xe Laser

This chapter marks the start of the experimental part of this thesis, beginning with a description of the apparatus and set-up used throughout the project before reporting on a number of important measurements made in the CW regime.

The experiment was originally built by Sam Bateman to record the data that is presented in Sect. 6.1.4 [1] but was subsequently stripped down, redesigned and rebuilt for the purpose of this work. The set-ups presented in Sects. 6.1.2 and 6.1.3 are largely the reconstructed versions, although for chronological reasons certain components that were used in the preliminary experiments before they underwent substantial changes (described in Sect. 6.1.5) are featured.

As will be presented in this chapter, early observations of the discharge appeared to show it behaving as a stable continuous discharge. It is later shown in Chap. 7 that this was not the case. The results in this chapter are still relevant however as they provide irrefutable evidence of the existence of gain within the discharges, a full analysis of the polarisation of the laser output and a demonstration of the long term stability and self-sufficiency of the system.

6.1 Experiment Assembly and Preliminary Results

6.1.1 Optical Fibre Selection

As discussed in Chap. 5 the free boundary anti-resonant design of fibre was the ideal candidate for this project with its low bend loss, large core to support a discharge and good IR guidance to exploit the high gain xenon laser lines.

The two fibres used throughout this thesis were 10 resonator free boundary fibres fabricated by Walter Belardi. Figure 6.1a is an SEM of fibre A, which has outer and inner diameters of 273 μm and 120 μm respectively. Fibre A suffers from losses no greater than 0.4 dB m^{-1} throughout its main guidance band between 3-4 μm as shown in Fig. 6.1b, with a measured loss of 0.16 dB m^{-1} at 3.5 μm. Fibre B is of a very similar design, but has an outer diameter of 264 μm.

© Springer International Publishing AG, part of Springer Nature 2018
A. Love, *Hollow Core Optical Fibre Based Gas Discharge Laser Systems*,
Springer Theses, https://doi.org/10.1007/978-3-319-93970-4_6

(a)   **(b)**

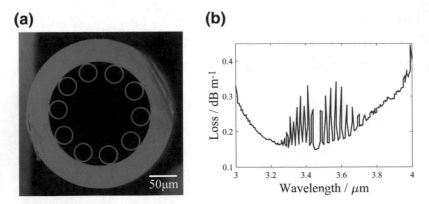

Fig. 6.1 **a** SEM of fibre A, with a minimum core diameter of 120 μm and resonators of diameter 35 μm and wall thickness 2.8 μm. **b** Measured optical attenuation of the same fibre recorded by Walter Belardi. The fine peaks between 3.2 and 3.8 μm are absorption peaks of HCl that are a remnant of the manufacturing process and can be removed by purging and evacuating the fibre [2]

The broadband transmission of both fibres can be seen in Fig. 6.2. For each fibre one trace was taken while the fibre was straight, and a second was taken with a bend of radius ≈ 20 cm—similar to that which would be experienced in later experiments. In both fibres the mid-IR band is negligibly affected by the bend with < 0.1 dB of loss.

6.1.2 Vacuum and Gas Handling System

Gas discharges are highly susceptible to impurities, with even small amounts of contaminants affecting the stability and optical output of a discharge. Isolating the system and applying a vacuum not only eliminates this, but also gives much greater control over the gas mixtures and pressures inside the fibre.

The vacuum system used for these experiments was constructed from a combination of KF, Swagelok and custom fittings, the configuration of which can be seen in Fig. 6.3. The system was evacuated by an œrlikon leybold Turbolab 80 vacuum pump, a two stage pump featuring a Turbovac SL80H turbo pump backed up by a dry diaphragm pump to reach a pressure of the order of 10^{-5} mbar in the system. The vacuum pressure was monitored by a penning gauge while the pressure in the fibre was monitored at both ends by capacitance gauges, which provide a gas-independent readings down to 10^{-2} mbar. The gases were supplied from bottles of helium (N5.5 purity), neon (N5.0 purity) and xenon (N5.0 purity).

The fibre was sealed in gas cells specially designed to allow access to the ends of the fibre. The gas cell that housed the grounded cathode end of the discharge can be seen in Fig. 6.4. The cell was made from a brass block that was milled to feature space for a 2 mm thick sapphire window and a removable fibre holder, then fitted

(a)

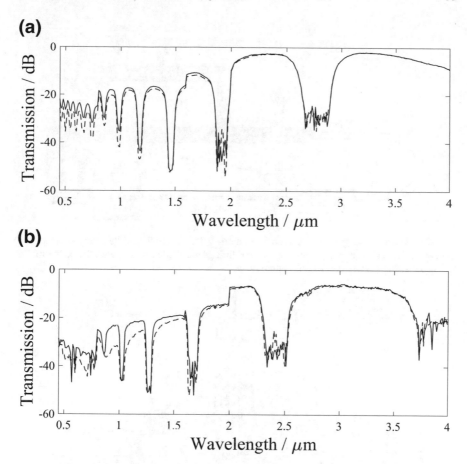

(b)

Fig. 6.2 Transmissions from a white light source through **a** 65 cm of fibre A and **b** 95 cm of fibre B for straight (solid) and bent (dashed) pieces of fibre. Measurements were taken using a Bentham DTMc300 Monochromator in combination with Si, InGaAs, PbSe and InSb detectors at a resolution of 10 nm. To remove the spectral variations in the detectors and halogen white light source the measurements were normalised to a separate spectra of the source. However there are still discontinuities at 800, 1600 and 2000 nm where the detectors change due to the differing active areas of the detectors

with a pipe for gas delivery and evacuation. The fibre could be sealed in the holder by a rubber bung with a drilled 300 μm hole compressed by a screw. The head of the fibre holder was also made of brass, and acted as the cathode electrode.

The high voltage anode end of the discharge required housing in an insulting material. For the preliminary experiments optical access to this end was not required, so a nylon Swagelok fitting was used. The fitting was customised so that a 1 mm thick tungsten electrode pin could be positioned close to the end of the fibre. A piece of plastic tubing was then used to connect the cell to the rest of the vacuum system so that it was isolated electronically.

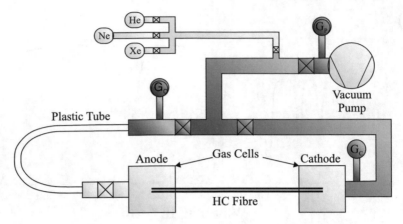

Fig. 6.3 Schematic diagram of the vacuum system. Valves are indicated by crosses, capacitance gauges by G_C and the penning gauge by G_p. KF fittings are shaded darker, with Swagelok fittings in lighter shading. The gas cells are custom made

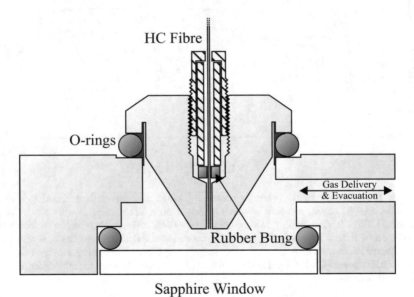

Fig. 6.4 Schematic diagram of the cathode gas cell. The majority of the cell (shaded light grey) is made of brass, while the rod and screw of the fibre holder (diagonal lined) are made of Delrin plastic. The fibre holder and window are secured in place by brass supports screwed in the cell

    Before being sealed in the gas cells, the fibre ends were cleaved to ensure that the end faces were clean and flat. As the lengths of all the fibres were measured prior to this, all quoted lengths to follow come with a ± 2.5 cm uncertainty.

    The time taken to fill a fibre with gas can be estimated by

$$\tau = \frac{128\,\eta\,l_t^2}{\pi^2\,d_t^2\,P_0},$$ (6.1)

which is the $\frac{1}{e}$ time to fill a closed tube of length l_t, diameter d_t and pressure difference P_0 with a gas of viscosity η [3]. Taking $l_t = 0.75$ m, half the longest length of fibre used as both ends are filled, $d_t = 120$ μm and $\eta \sim 2 \times 10^{-5}$ Pa s [4], the time taken to fill the fibre to a pressure of 12 mbar will be ~ 10 s.

However as Eq. 6.1 is based on the fluid dynamics of a viscous, laminar flow it will only be applicable if the mean free path of the gas is much less than the tube diameter. The mean free paths of helium and xenon were calculated from Eq. 3.2 to be ~ 200 μm and ~ 15 μm respectively at 12 mbar, and so comparable to the fibre ID. This means that any value calculated from Eq. 6.1 will be incorrect, however it will provide a reasonable estimate allowing it to be assumed that the fibre will be filled in at most a matter of minutes.

Once filled, the valves around the fibre were closed so that the rest of the system could be evacuated before the voltage was turned on. This was to avoid any unwanted discharging happening within the system itself, particularly through the plastic tubing connected to the anode cell.

The time it takes to evacuate the fibre is more difficult to estimate. In the initial stages of evacuation the rate will be similar to when filling, but with a reducing pressure the mean free path of the remaining gas particles rapidly increases. This will push the fluid dynamics far out of the viscous flow regime and into free molecular flow (otherwise known as Knudsen flow [5]), where the flow rate is substantially lower and evacuation takes an extremely long time. To combat this and remove any impurities trapped in the fibre, the system was flushed with helium multiple times before use.

6.1.3   Electrical System

The electrical set-up can be seen in Fig. 6.5. The potential difference required to achieve breakdown was supplied by an UltraVolt 40A24-P30 power supply, capable of producing a variable voltage of up to 40 kV with a maximum current of 0.75 mA. The supply featured output ports through which the voltage and current drawn from the supply could be monitored.

The power supply was connected to the anode via a ballast resistance constructed of seventy seven 1 MΩ resistors connected in series and submerged in a polyurethane potting compound to avoid unwanted arcing between resistors.

The voltage across the discharge was measured by a Tenma 72-3040 high voltage probe, which was rated up to 40 kV and had a load resistance of 1 GΩ. The supply, probe and cathode were then all connected to the ground to complete the circuit. The current, supply voltage and discharge voltage were all measured by Tenma 72-7735

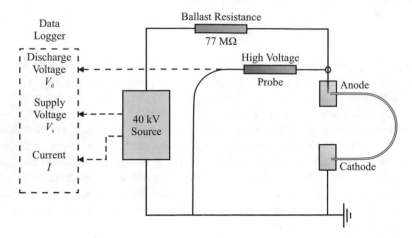

Fig. 6.5 Schematic diagram of the electrical system

multimeters, with their data values recorded at 1 s intervals by an accompanying piece of data logging software.

6.1.4 Preliminary Experiments

As stated in the introduction to this chapter, this section summarises preliminary work carried out by Sam Bateman. For the reasons outlined in Sect. 4.3 a helium-xenon mixture was chosen as a starting point for this experiment, while the chosen pressures were based on the work by Smith and Maloney [6]. In their work it was demonstrated that a 5 : 1 mixture of ^{3}He and ^{136}Xe at a pressure of around 6.5 torr (8.7 mbar) yields a very high gain in a tube with an inner diameter of 250 μm. In an attempt to keep the pD ratio roughly the same, a 5 : 1 ratio of He-Xe at a pressure of 12 mbar was used—the slight discrepancy arising from the differing isotopes of helium and xenon used.

These early discharges proved to be successful in lengths of fibre up to 1 m in length. The signal was analysed using a Bentham TMc300 monochromator (featuring a 300 grooves mm^{-1} grating and variable filters) with a cryogenically cooled InSb detector (with a spectral range of 1 to 5 μm and peak responsivity of 3 A W^{-1} at 3.8 μm), a Bentham 477 current pre-amplifier (with a gain variable from 10^3 to 10^8) and a Bentham 496 lock-in amplifier (time constant 10 ms to 10 s).

Figure 6.6a is a spectrum collected from the cathode end of a 50 cm long discharge in which a number of xenon lines can be observed, with the 3.5 μm line being particularly strong.

An analysis of the intensity of five individual lines for discharge lengths of 50, 80 and 100 cm is presented in Fig. 6.6b. The intensities of the 3.11, 3.36 and 3.51 μm

(a)

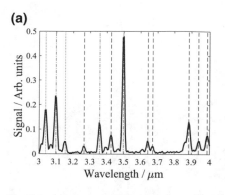

(b)

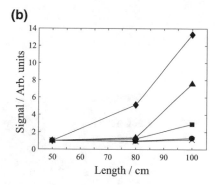

Fig. 6.6 **a** Spectra of a helium-xenon gas discharge inside 50 cm of fibre A at a 5 : 1 ratio of He to Xe at a pressure of 12 mbar, taken with a 5 nm resolution. The dotted, dashed and dot-dashed lines identify known xenon lines that do not lase, have demonstrated lasing and have shown evidence of gain in this system respectively. **b** Line analysis of the 3.04 (crosses), 3.11 (squares), 3.36 (triangles), 3.51 (diamonds) and 3.89 μm (circles) lines. *Data taken from* [1]

lines all demonstrate a super-linear growth with respect to discharge length, a strong indication that these lines are experiencing optical gain.

Conversely, the 3.04 and 3.89 μm lines, the latter of which shares an upper energy level with the 3.51 μm line, do not shown any real growth. This implies that the observed signal increase is not simply due to a fluorescence effect, which would affect both lines an equal amount, but is instead ASE.

6.1.5 Modifications

With strong indications of gain present on at least three xenon lines, the next step to constructing a laser was to build a cavity. This however required significant modifications to the set-up.

The anode gas cell underwent the largest change, with a completely new cell constructed. As plastics have a tendency to adsorb gases then outgas when under vacuum, the new cell was made out of Macor ceramic. However the brittle nature of ceramic meant that the new cell could not just be a copy of the existing metal cell.

The final design can be seen in Fig. 6.7. The fibre was sealed in the same way as before but with a small o-ring replacing the rubber bung. A 5 mm thick calcium fluoride window was positioned opposite the fibre holder to provide optical access and a custom brass fitting was inserted perpendicular to the optical axis to connect the cell to the vacuum system and to hold the anode pin.

The cathode cell also received a slight modification after it was discovered that the discharge would often punch a hole through the fibre in the small region between the end of the fibre holder head and the rubber bung. The change can be seen in Fig. 6.8, where the hole has been enlarged to a diameter of 1 mm so that an extra

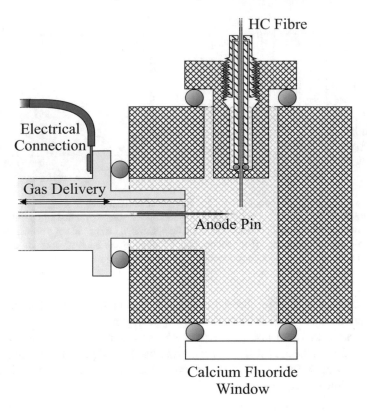

Fig. 6.7 Schematic diagram of the modified anode gas cell. The main body of the cell (hashed) is made of Macor ceramic, the gas delivery tube (shaded light grey) is brass, the rod and screw of the fibre holder (diagonal lined) are Delrin plastic and the anode pin is tungsten. The window, gas tube and fibre holder are all secured in place by Delrin plastic fittings screwed into the cell (not pictured)

glass tube (820 μm OD 287 μm ID) could be inserted to provide extra insulation. The fibre was sealed in the tube by a vacuum adhesive (Torr Seal), and the outside of the tube was then sealed by an o-ring.

The electronics were left mostly unchanged, however the fixed ballast resistance was replaced by a variable one. The ballast was still constructed of 1 MΩ resistors connected in series and submerged in a potting compound, but now the resistors were tapped in a way such that a resistance of any multiple of 6 MΩ between 6 and 78 MΩ could be achieved by using different connecting ports. This gave much greater control over the current drawn from the supply.

Fig. 6.8 Schematic diagram
of the modified cathode gas
cell fibre holder

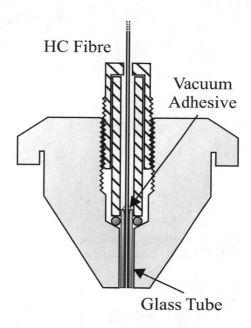

HC Fibre

Vacuum
Adhesive

Glass Tube

6.1.6 Optical Set-Up

To minimise the number of degrees of freedom in the alignment optomechanics, the
laser cavity was formed simply by using a pair of silver spherical mirrors in front of the
fibre ends as seen in Fig. 6.9. Mirrors with a 100 mm radius of curvature (ThorLabs
CM254-050-P01) were chosen so that additional optics like a calcium fluoride 50/50
beamsplitter for 3.5 μm (ThorLabs BSW511) could be placed between them and the
fibre ends as required. The mirrors were placed on translation stages to provide fine
control of the focal length.

The beamsplitter acts as an output coupler to reflect a fraction of the signal off
for analysis by a cryogenically cooled InSb detector, connected to a Stanford SR830
lock-in amplifier. In an attempt to isolate the 3.5 μm xenon line of most interest a
bandpass filter (ThosLabs FB3500-500) was placed in front of the detector, however
with a bandwidth of 500 nm the 3.36 μm line will also be detected.

With such a large gap between the fibre ends and the mirror it was necessary to
calculate the width of the beam at the mirror position so that the mirror would be
large enough to collect all the light. The beam width can be calculated from Gaussian
optics,

$$w(z) = w_0\sqrt{1 + \left(\frac{z}{z_r}\right)^2},$$ (6.2)

where w_0 is the beam waist and

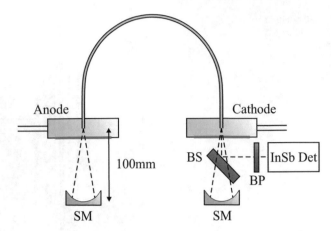

Fig. 6.9 Schematic diagram of the optical set-up; SM - silver coated spherical mirror, BS - CaF$_2$ beamsplitter, BP - 3.5 μm bandpass filter

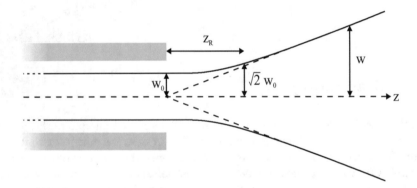

Fig. 6.10 Divergence of a Gaussian beam exiting a hollow core fibre

$$z_r = \frac{\pi w_0^2}{\lambda},\qquad (6.3)$$

is the Rayleigh length, defined as the distance where $w(z_r) = \sqrt{2}w_0$. For a beam exiting a hollow optical fibre the beam waist is equal to the mode radius, and so its profile can be calculated as in Fig. 6.10. Based on modelling by Walter Belardi, the fundamental mode radius of fibre A is 40 μm which gives a 1.4 mm Rayleigh length for a wavelength of 3.5 μm. This then gives a beam width of 5.6 mm at a point 100 mm from the fibre, and so a standard 25.4 mm diameter mirror is more than sufficient.

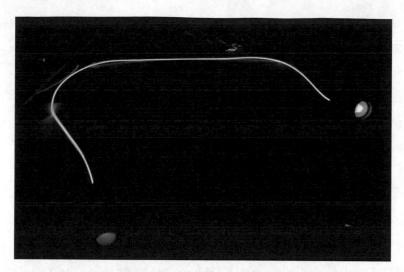

Fig. 6.11 Photograph of a helium-xenon discharge 1 m in length. The high voltage anode glow is on the left, with the cathode glow on the right

6.2 Discharge Operation

With the new gas cells in place, stable discharges of up to 1.5 m in length were now attainable in a 5 : 1 helium-xenon mixture at 12 mbar. Figure 6.11 shows a photograph of a stable discharge in a 1 m long piece of fibre A. The discharges were bright blue in colour, as would be expected in a xenon discharge, with a purple glow emanating from the cathode. From the geometry of the cathode it is believed that this purple glow is some combination of the cathode and negative glows which are always perpendicular to the cathode surface, with the positive column forming the vast majority of the discharge in the fibre.

The electrical properties of a typical discharge can be seen in Fig. 6.12. Aside from a settling down period during the first 100 s, there is no real degradation of the discharge from cataphoretic effects or leakages over the 10 minutes for which the discharge is operating—with later discharges remaining stable for up to and over 30 minutes.

Figure 6.13 is a current-voltage plot for a similar discharge where the supply voltage was stepped up and down. Prior to breakdown, the supply and discharge voltages are closely matched, with only a small leakage current being drawn. Upon reaching 27.7 kV the discharge breaks down with V_d dropping to 18.9 kV and I increasing sharply to 0.2 mA. When V_s is further increased I also increases and V_d decreases, suggesting that the discharge is operating in the normal glow regime. On decreasing V_s the discharge then displays hysteretic behaviour, extinguishing at a lower voltage than at breakdown. All these observations combined gave no reason to believe that the discharges were anything but continuous.

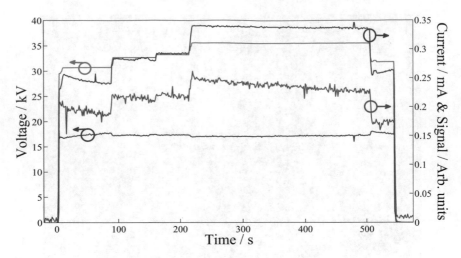

Fig. 6.12 Electrical properties of a typical 5 : 1 helium-xenon gas discharge at a pressure of 12 mbar inside 70 cm of fibre A with a ballast resistance of 54 MΩ. Left axis; supply voltage (green) and discharge voltage (blue). Right axis; current (red) and signal (purple)

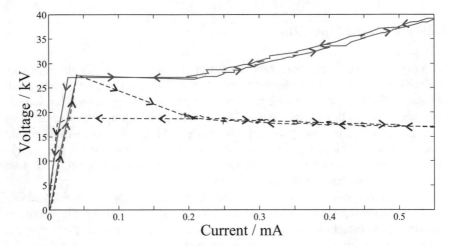

Fig. 6.13 Current-voltage plot for the supply (solid) and discharge (dashed) voltages of a 5 : 1 helium-xenon gas discharge at a pressure of 12 mbar inside 90 cm of fibre A with a ballast resistance of 42 MΩ. One data point was recorded every second

    While individual discharges suffered from minimal degradation over their life-
time, continual discharges did degrade the fibres themselves. This effect was particularly apparent at the fibre ends, as can be seen in the optical micrographs taken
of both ends after a significant number of discharges in Fig. 6.14. The cathode end
(Fig. 6.14a) is relatively unaffected structurally, but there appear to be bands of colour
around the outside probably stemming from thin layers of material deposition that

(a) **(b)**

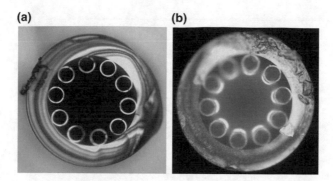

Fig. 6.14 Micrographs of the **a** cathode and **b** anode ends of a fibre after a number of discharges

reflect light much like the anti-resonant reflections that cause the fibre to guide. This deposition is also present on the sapphire window in front of the fibre, requiring it to be replaced after prolonged use. Conversely the anode end (Fig. 6.14b) shows more significant damage to both the resonators and the jacket, and it was this end which would often fail first.

6.3 Double-Pass Experiments

For the first set of experiments only the mirror in front of the anode was used and the signal from the cathode cell was sent straight into the Bentham spectrometer. Such a set-up is referred to as a *double-pass* as the light is made to pass through the gain medium twice.

The alignment of invisible 3 μm optics proved to be a challenge, however the achromatic nature of the spherical mirror meant that the visible light generated by the discharge could be used as a rough guide to position the mirror. The coupling could then be optimised by making fine adjustments while monitoring the infrared signal output.

Once the mirror was aligned the signal could be blocked by simply placing a piece of card in front of the mirror, providing an easy and repeatable way of comparing the signal from single- and double-pass measurements.

Figure 6.15 shows one such measurement in a 90 cm long piece of fibre A. The current was increased in steps of 0.05 mA from 0.2 to 0.55 mA then reduced in a similar fashion back to 0.2 mA, with single- and double-pass measurements taken at each step. A ballast resistance of 42 MΩ was used to achieve this range of current. The data was then normalised to the first single-pass measurement at 0.2 mA.

When feeding light into an excited medium, a strong increase of signal is a sign that more stimulated emission is occurring than absorption and the medium is exhibiting gain. These results show an increase in signal between the two data sets by a factor that is consistently greater than 2 which, in light of the fact that a roughly even signal

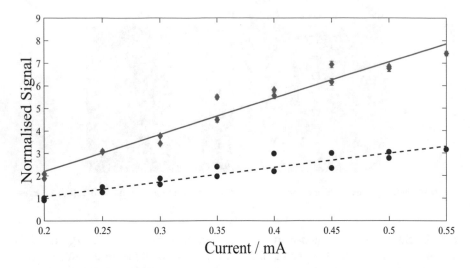

Fig. 6.15 Signal at various currents for single-pass (dashed) and double-pass (solid) discharges in 90 cm of fibre A with a 5:1 ratio of helium-xenon at a pressure of 12 mbar and a 42 MΩ ballast resistance. The measured signals have been normalised to the initial 0.2 mA single-pass reading

was detected at both ends, not only confirms the previously suggested presence of ASE, but also indicates a high level of gain. The linear signal-current trend is revisited in Chap. 7.

The spectra in Fig. 6.16 were taken from the same set-up at a current of 0.25 mA after the fibre was refilled with an equivalent gas mixture. The signal is again normalised to the initial 3.5 μm single-pass reading. While the 3.5 μm line still shows evidence of double pass ASE, the 3.1 and 3.36 μm lines do not, which suggests that absorption is occurring on these lines.

Figure 6.17 is the same measurement as Fig. 6.15 repeated in a fresh, 1 m long piece of fibre A. The similarities between the two measurements imply that this behaviour is very much repeatable.

6.4 Completing the Cavity

With the alignment of both mirrors as in Fig. 6.9, the laser cavity is completed. To characterise the behaviour of the cavity and confirm lasing the mirrors were blocked in a similar fashion to the double-pass experiments, however in this case the two mirrors provide four possible combinations. To avoid confusion a naming convention which is outlined in Table 6.1 was adopted.

Figure 6.18 is a snapshot of the signal during a discharge in a 1.5 m piece of fibre B with a current of 0.25 mA, during which the mirrors were sequentially blocked and revealed. When going from single-pass to reverse double-pass, the resulting double-

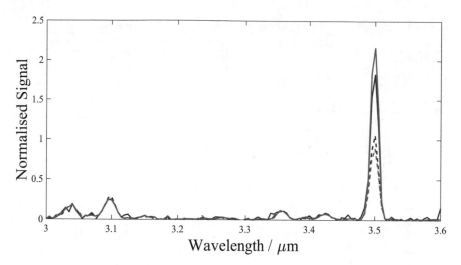

Fig. 6.16 Spectra for single-pass (dashed) and double-pass (solid) discharges in 90 cm of fibre A with a 5 : 1 ratio of helium-xenon at a pressure of 12 mbar, a discharge current of 0.25 mA, a 42 MΩ ballast resistance and a 20 nm resolution

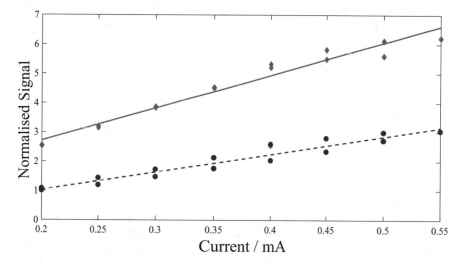

Fig. 6.17 Repeat of the double pass measurement in a fresh 1 m length of fibre A. All other parameters are the same as in Fig. 6.15

pass ASE was predominantly backwards-going, and the reduced signal observed indicates that the gain becomes saturated. The transition from single-pass to forward double-pas behaves in the same fashion as before.

Table 6.1 Naming convention for analysis of a cavity with a pick off at the cathode

Name	Abbreviation	Anode mirror	Cathode mirror
Single-Pass	SP	✗	✗
Forward Double-Pass	FDP	✓	✗
Cavity	C	✓	✓
Reverse Double-Pass	RDP	✗	✓

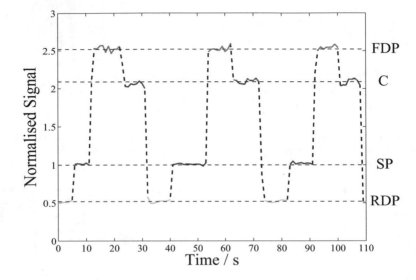

Fig. 6.18 Monitored output signal (black dotted line) from a 5:1 helium-xenon discharge at a pressure of 12 mbar inside 1.5 m of fibre B as the cavity mirrors were sequentially blocked and revealed. The signal has been normalised to the average SP reading. The signal levels in the different cavity configurations have been highlighted for clarity (SP - blue, FDP- green, C - red, RDP - cyan)

As the output beam is from the mode of the optical fibre and the gain is saturated on just a double-pass, little change in power and no change in beam profile is expected when the cavity is formed to make a laser instead of just ASE.

A significant drawback of using a CaF_2 beamsplitter as an output coupler is that the reflections suffer from a polarisation dependence, with light in the S- and P-polarisations having reflection coefficients of 0.6 and 0.3 respectively at 3.5 μm. Provided that the polarisation states of a laser cavity are well defined, a lower cavity loss on one polarisation results in that polarisation consuming more of the available gain and hence a laser that is preferentially in that polarisation.

Despite possessing no intrinsic birefringence, the lack of any material stress-induced refractive index perturbations in a hollow core fibre results in a minimal amount of crosstalk between the two polarisation modes. The geometry of the fibre will however cause the polarisation to rotate in the same way that a Fresnel rhomb will [7].

If the path of an optical fibre is defined by a space curve r then at any point P along the fibre there exists an orthonormal basis (t, n, b), where t is the tangent, n is the normal pointing towards the centre of curvature, and b is the binormal perpendicular to the (t, n) plane (also referred to as the osculating plane). The direction of the state of polarisation of light in the fibre can then be given by

$$E = E_n n + E_b b, \tag{6.4}$$

where $E_{n,b}$ are the components in the n and b directions. The change of E along the fibre between two points separated by a path length of ds is [8, 9]

$$\frac{dE}{ds} = \begin{bmatrix} 0 & -\tau \\ \tau & 0 \end{bmatrix} E \tag{6.5}$$

where τ is the torsion of the curve, a measure of how fast b rotates. Equation 6.5 describes a rotation of the polarisation by an amount τ, matching the rotation of the osculating plane. The total rotation along the fibre can be calculated by integrating τds over the length of the curve. Note that if the fibre is in a plane, $\tau = 0$ and the polarisation angle is unchanged.

This polarisation rotation can be observed in Fig. 6.19, where the signal from the anode was monitored using a polariser and detector while the spherical mirror and beamsplitter at the cathode end were set up as in Fig. 6.9. The double-pass signal was recorded at 10° intervals of the polariser, then normalised to the single-pass measurement at each point. The polarisation dependence from the CaF$_2$ beamsplitter

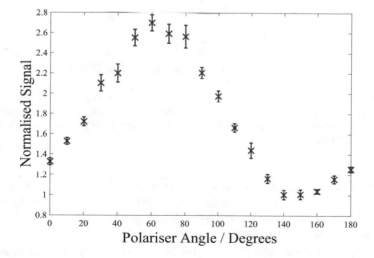

Fig. 6.19 Double-pass signal from the anode end of a 5:1 helium-xenon discharge at a pressure of 12 mbar inside 1.5 m of fibre B for various angles of an output polariser. The signal has been normalised to the single-pass signal. The CaF$_2$ beamsplitter placed between the cathode end and mirror creates a polarisation dependence which is rotated by the geometry of the fibre

(a)

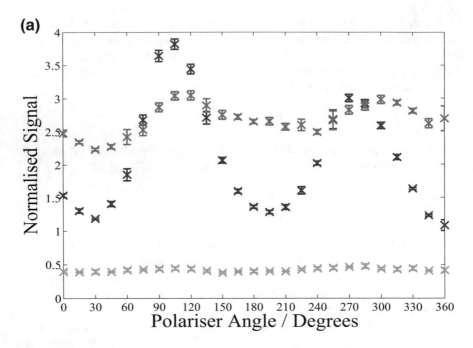

(b)

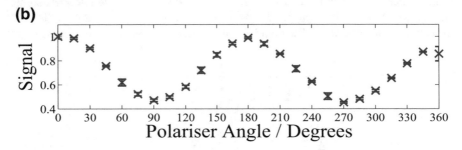

Fig. 6.20 Cathode output signals through a polariser from a 5:1 helium-xenon discharge at a pressure of 12 mbar inside 1.5 m of fibre B. **a** shows the forward double-pass (green), cavity (red) and reverse double-pass (cyan) cavity configuration output signals which have been normalised to the single-pass output signal shown in (b)

is very clear, and as the polariser has been calibrated so that the S- and P-polarisations line up with 0 and 90°, the fibre has induced a $\approx 30°$ rotation of the polarisation.

Fortunately, as this rotation is purely due to the fibre geometry it should be reversible. Torsion is directionally dependent, so τ is replaced with $-\tau$ in Eq. 6.5 when going in the reverse direction through the fibre and so the polarisation angle is restored to its original value.

This means that the cavity polarisations are well defined and the cavity can preferentially lase in the output P-polarisation. Figure 6.20 shows the output signals from the forward double-pass, cavity and reverse double-pass arrangements of the cavity, normalised to the single-pass signals, at various angles of a polariser.

While the forward and reverse double-pass signals are relatively flat, the cavity signal shows a strong signal increase for the P-polarisation. The clear polarisation dependence is evidence that multiple round trips are occurring within the cavity, providing the necessary feedback for the system to be described as a laser, not simply ASE.

The reduced signal in the second peak is likely to be an ageing affect of the discharge. There also appears to be a slight shift in angle from the S- and P-polarisation for first maxima and minima in the cavity signal of around 10° to 30°. The reasons for this are not clear, particularly as the second maxima does line up with the P-polarisation.

6.5 Summary

In this chapter DC excited glow discharges in a hollow core fibre have been demonstrated. Using a 5:1 mixture of helium-xenon at a pressure of 12 mbar inside 1.5 cm of a 120 μm ID hollow core free-boundary fibre, electronic breakdown was achieved at potential differences of < 40 kV and could be sustained for well over ten minutes. The discharges exhibited all the signs of a continuous normal glow discharge.

Previous work [1] had shown that the signal on the 3.11, 3.36 and 3.51 μm lines from such a discharge exhibited gain. This was verified in a double-pass experiment, where the output signal from the anode end of the discharge was reflected back into the fibre by a spherical mirror. The output signal from the cathode end increased by a factor greater than two on the 3.51 μm confirming gain, however no change was observed on the other two lines.

A complete cavity was then constructed around the fibre, using a CaF_2 beamsplitter as an output coupler at the cathode. A reduction in the output signal when a mirror was aligned to reflect light back through the output coupler indicated that gain saturation was occurring from a double-pass alone. A polarisation dependence in the cavity output signal caused by 60 and 30% reflections in the S- and P-polarisations from the output coupler showed that oscillation was occurring in the cavity. This confirmed the construction of a laser.

References

1. S. A. Bateman, "Hollow core fibre-based gas discharge laser systems and deuterium loading of photonic crystal fibres," Ph.D. thesis, University of Bath (2014)
2. N.V. Wheeler, A.M. Heidt, N.K. Baddela, E.N. Fokoua, J.R. Hayes, S.R. Sandoghchi, F. Poletti, M.N. Petrovich, D.J. Richardson, Low-loss and low-bend-sensitivity mid-infrared guidance in a hollow-core-photonic-bandgap fiber. Opt. Lett. **39**(2), 295 (2014)
3. D. Tabor, *Gases* (Penguin Books, Liquids and Solids, 1969)

4. D. R. Lide, *CRC Handbook of Chemistry and Physics*, CRC Press (2008)
5. M. Rieutord, *Fluid Dynamics: An Introduction*, Springer (2015)
6. P.W. Smith, P.J. Maloney, A self-stabilized 3.5-μm waveuide He-Xe laser. Appl. Phys. Lett. **22**, 667 (1973)
7. J. Strong, *Concepts of Classical Optics* (Dover Books on Physics, Dover Publications, 2012)
8. J.N. Ross, The rotation of the polarization in low birefringence monomode optical fibres due to geometric effects. Opt. Quantum Electr. **16**, 455 (1984)
9. E.M. Frins, W. Dultz, Rotation of the polarization plane in optical fibers. J. Lightwave Technol. **15**, 144 (1997)

Chapter 7
Pulsed Measurements of the He-Xe Laser

Following some fortunate measurements made by the author it was discovered that rather than a continuous discharge, the system was in fact operating in a semi-stable pulsing regime with a strong mid-IR afterglow light pulse. This led to measurements with a more favourable signal-to-noise ratio, and so a simpler alignment.

This chapter presents all the experiments performed with helium-xenon discharges measured in this pulsed regime. After characterising the pulsing, the laser cavity was analysed for its polarisation and longitudinal mode beating properties before a number of pressure optimisation experiments were carried out.

7.1 A New Measurement Regime

7.1.1 Electrical Properties

In an attempt to investigate the discharge behaviour a little more and hopefully improve the signal-to-noise ratio making the mirror alignment easier, the discharge voltage probe was connected to an oscilloscope. An example of the resulting time traces can be seen in Fig. 7.1.

The voltage is seen to be sharply oscillating with a saw-tooth profile not unlike those of a Pearson-Anson oscillator described in Sect. 3.5. In this case however the capacitance required to cause this pulsing behaviour is coming from the anode gas cell surroundings and the metal value connected to the cell; it is stray capacitance.

The value of the stray capacitance can be calculated from the heights of these pulses by using Eq. 3.18. Using values of T, R, V_s, V_{br} and V_{ex} from a number of different discharges, C is found to lie between 10 and 20 pF.

The inclusion of a capacitor in the circuit means that the previously recorded current is actually flowing to the capacitor, not through the discharge. It then follows that the linear relationship between the measured current and signal observed in the double-pass measurement of Fig. 6.15 was a result of the capacitor charging rate and subsequent pulse repetition rate, not the discharge behaviour. The current flow

© Springer International Publishing AG, part of Springer Nature 2018
A. Love, *Hollow Core Optical Fibre Based Gas Discharge Laser Systems*,
Springer Theses, https://doi.org/10.1007/978-3-319-93970-4_7

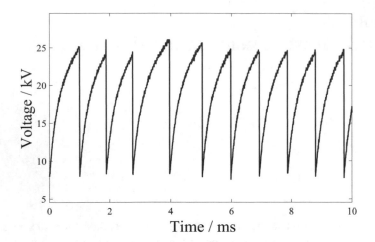

Fig. 7.1 Time trace of the voltage across a discharge

was increased by increasing the supply voltage, which in turn charged the capacitor faster leading to more frequent pulses and a higher average signal.

Figure 7.2 plots the CW measured single-pass discharge output signal and pulse repetition rate against the recorded current. The pulse frequency was calculated from the Fourier transform of a 2 s window of the pulsing discharge voltage. Linear trendlines have been fitted for each data series, with gradients of 4.5 ± 0.2 AU mA^{-1} for the signal and 3.3 ± 0.1 kHz mA^{-1} for the repetition rate. These show that a correlation exits between the two variables, but also suggest that there may also be some other factor as the gradients fall outside of each others error bounds.

While the pulsing neatly explains the signal-current trend, it unfortunately also means that the actual current flowing through the pulsing discharge is essentially unknown. If the voltage drop across the discharge with respect to time is known however, the current can be estimated by

$$I = I_s + I_C = I_s - \frac{dV}{dt}C, \tag{7.1}$$

where I_s and I_C are the currents from the supply and capacitor respectively. The voltage probe used previously in Chap. 6 can be used to measure the discharge voltage as it strikes, but these measurements must be taken with a certain degree of scepticism as the probe was designed for DC applications.

The dotted curve in Fig. 7.3 shows the discharge voltage as measured by the Tenma 72-3040 probe with a Keysight DSOX3032A oscilloscope set to an averaging of 128. The ringing feature was suspected to be an artefact of the probe and not a real measurement. A Testec TT-HVP 15HF probe with a bandwidth of 50 MHz was trialled to get a faster, more accurate reading. However with a load resistance of only

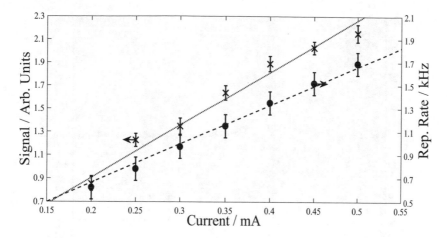

Fig. 7.2 Variation of average signal (crosses, left axis) and voltage pulse frequency (circles, right axis) with the measured current drawn from the power supply for 5:1 helium-xenon discharges at a pressure of 12 mbar inside 70 cm of fibre B. Trendlines with gradients of 4.5 ± 0.2 AU mA^{-1} for the signal (dotted) and 3.3 ± 0.1 kHz mA^{-1} for the repetition rate (dashed) have been fitted so that they pass through the origin

100 MΩ (comparable to the ballast resistance) the Testec probe was too much of a drain on the system and the discharge would not strike.

The dashed curve in Fig. 7.3 shows this voltage after applying a numerical low-pass filter to remove the ringing. The smoothed voltage was then differentiated and used in combination with a capacitance of 14 pF, calculated from a separate measurement of this set of discharges, to estimate the discharge current, plotted as a solid line.

With a peak current density of ≈ 4000 A cm^{-2} this pulsing is not only putting substantially more strain on the fibres than thought, but also could be pushing the discharge into an arc-like behavioural regime. The latter point seems unlikely however due to the lack of green colouration in the cathode glow from the copper within the electrode (a characteristic feature of arc discharges).

Figure 7.3 also shows a large negative current spike at the onset of the pulse. This is possibly another artefact of the slow voltage probe or it could be due to an inductance generated by the discharge. As the inductance of a discharge has been shown to increase with electrode separation [1], it is expected to be significant, although without more sophisticated equipment no conclusions can be drawn.

The high current pulse does however open the possibility of building an ion laser. To investigate this, a number of 1 nm resolution spectra of the discharge were taken through the visible and near-IR spectral regions in an attempt to identify the lines present. Two of the spectra can be seen in Fig. 7.4, while a number of the lines observed are presented in Table 7.1. Many more neutral xenon lines were observed beyond 1 μm, however these have not been included.

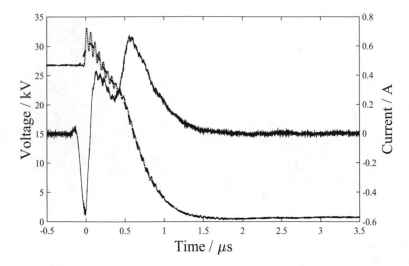

Fig. 7.3 Voltage and current across 5 : 1 helium-xenon discharges at 12 mbar inside fibre B averaged over 128 pulses. The dotted and dashed curves (left axis) represent the voltage before and after numerical low-pass filtering. The solid curve (right axis) represents an estimate of the current calculated from the dashed curve by Eq. 7.1

As the values for relative line strengths in Table 7.1 come from different experiments they cannot be used to directly compare the abundances of neutral and singly ionised xenon. However as they are of comparable strength it is reasonable to assume that there are enough excited xenon ions to be significant. The subject of an ion laser is revisited in the latter half of Chap. 8.

7.1.2 Optical Properties

As the electrical pulse in Fig. 7.3 is of the order 1 μs in duration, a new optical detector with a bandwidth of at least 1 MHz would be required to make accurate measurements. Two detectors were used initially; one for the near-IR observed earlier in Fig. 7.4b to quickly verify the existence of light pulses, and one for the mid-IR laser light. It was soon observed that the high voltage pulses were causing considerable interference in the detectors and that the detection system would have to be separated from the main set-up.

This was achieved through the use of a indium fluoride fibre patch cord (ThorLabs MF12, 100 μm core diameter, ≤ 0.45 dB m^{-1} attenuation from 2 to 4.6 μm) for mid-IR detection or a standard 600 μm core diameter fibre patch cord for visible to near-IR detection. Light was coupled into the patch cord by a 50 mm focal length CaF$_2$ lens (Thorlabs LA5183-E, or a fused silica lens for visible to near-IR) while the output was butt-coupled to the detector, as shown in Fig. 7.5.

The detector output was connected to the oscilloscope along with the discharge voltage probe, which was used as the waveform's trigger. The responsivities and gain

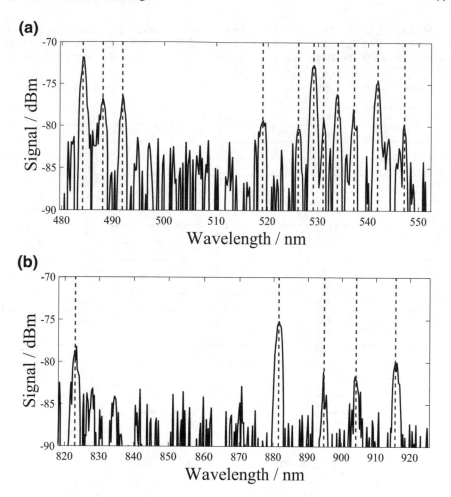

Fig. 7.4 Spectra of the light collected using a standard 600 μm core step-index multimode fibre patch cord (NA ≈0.4) from the side of a 5:1 helium-xenon discharge inside fibre B at a pressure of 12 mbar, measured by a Yokogawa AQ-6315A optical spectrum analyser with a resolution of 1 nm. Identified emission lines are marked with dashed lines

of the detectors were used to convert the oscilloscope output voltage into optical power. Where the exact wavelengths of the emission lines detected are unknown the peak detector responsivity has been used to provide a minimum estimate of power (this is mainly on the near-IR lines).

The initial measurements from a discharge in a 70 cm long piece of fibre B with both mirrors blocked can be seen in Fig. 7.6. The traces presented are the result of subtracting a 'dark' trace taken with the discharge on but the with the detector blocked from a 'light' trace.

Table 7.1 A selection of strongly observed lines in a 5:1 helium-xenon discharge inside fibre B at a pressure of 12 mbar as measured by a Yokogawa AQ-6315A optical spectrum analyser with a resolution of 1 nm. The accepted values, species and strengths are taken from [2] except for entries marked with * which come from [3]. ^{d} Denotes a doublet

Observed line/nm	Accepted wavelength/nm	Specie	Relative strength
484	484.433	Xe II	2000*
488	488.353	Xe II	600*
492	492.148	Xe II	800*
519	519^d	Xe II	400
526	526^d	Xe II	500
529	529.222	Xe II	2000
531	531.387	Xe II	1000
533	533.933	Xe II	2000
537	537.239	Xe II	500
542	541.915	Xe II	3000
547	547.261	Xe II	1000
823	823.164	Xe I	10,000
882	881.941	Xe I	5000
895	895.225	Xe I	1000
904	904.545	Xe I	400
916	916.265	Xe I	500
980	979.970	Xe I	2000
992	992.319	Xe I	3000

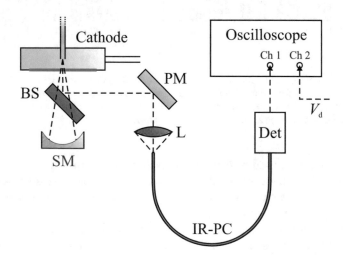

Fig. 7.5 Schematic diagram of revised cathode optical set up for the collection of light pulses. The rest of the system is left as in Fig. 6.9. PM - planar silver mirror, L - 50 mm focal length CaF$_2$ lens, IR-PC - InF$_3$ infrared fibre patch cord

(a)

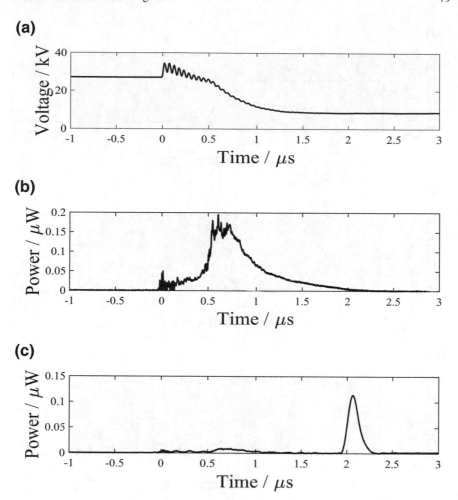

(b)

(c)

Fig. 7.6 Oscilloscope traces averaged over 128 pulses of 5∶1 helium-xenon discharges at 12 mbar inside 70 cm of fibre B showing **a** the discharge voltage, and the single-pass output power as was measured by **b** an InGaAs detector after amplification and **c** the signal from a fast mid-IR detector

Figure 7.6b was taken using a biased InGaAs detector (ThorLabs DET10C, 0.9 to 1.7 μm spectral range, 10 ns rise time, 1 A W^{-1} responsivity at 1.5 μm) in combination with a Laser Components HSA-X-2-40 high speed amplifier (2 GHz bandwidth, 5 kV A^{-1} gain), while Fig. 7.6b was taken using a VIGO PVI-3.4 IR photovoltaic detector (≤ 2 ns response time, 2.9 to 4 μm spectral range, 2.5 A W^{-1} responsivity at 3.5 μm) with a VIGO SIP-100-250M-TO39-NG preamplifier (10 kV A^{-1} gain).

The near-IR traces behave as predicted, with a ≈ 0.5 μs light pulse forming as the current reaches its peak, however the mid-IR traces feature a very strong pulse ≈ 0.5 μs after the current pulse has subsided in addition to a very faint pulse at the

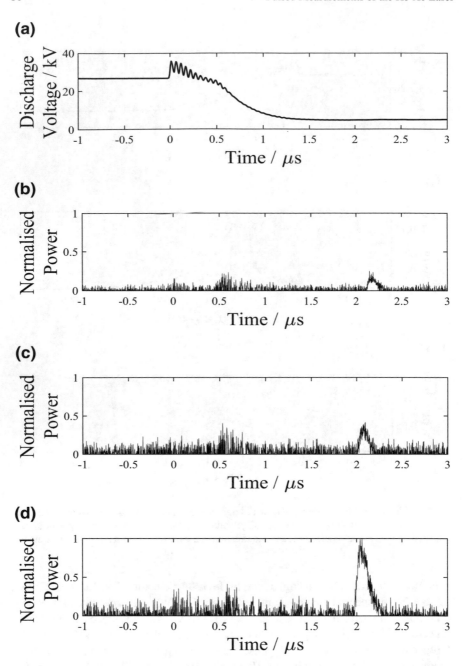

Fig. 7.7 Oscilloscope traces averaged over 64 pulses of 5:1 helium-xenon discharges at 12 mbar inside 70 cm of fibre B showing **a** the discharge voltage and the single-pass output powers measured through the Bentham TMc300 monochromator at **b** 3.11 μm, **c** 3.36 μm and **d** 3.51 μm. The power traces have been normalised to the 3.5 μm line peak power. The 3.11 and 3.36 μm measurements were made with a 20 nm resolution, while the 3.51 μm measurement was made with a 10 nm resolution

current peak. The presence of this strong afterglow pulse is perhaps not too surprising considering the high concentration of xenon ions suggested in the previous section, and based on the short duration (≈ 200 ns FWHM) of the pulse it is reasonable to assume that its presence is mainly fuelled by electron-ion recombination.

With an spectral range of 2.9 to 4 μm, a spectral measurement was required to verify that the light measured by the IR detector was at the expected wavelengths. Figure 7.7 shows measurements made with the fast IR detector through the Bentham TMc300 monochromator at 3.11, 3.36 and 3.51 μm respectively. They confirm that the light in Fig. 7.6c is primarily comprised of emission from the 3.51 μm line.

7.2   Cavity Analysis

7.2.1   Polarisation Test

On completion of the cavity, the near-IR lines detected by the InGaAs detector undergo no change whilst the mid-IR afterglow exhibits a reaction very similar to the CW signal measured in Chap. 6. However the more favourable signal-to-noise ratio did make for a much easier mirror alignment, which in turn made for a better coupling. Figure 7.8 demonstrates how the (b) unpolarised, (c) S-polarised and (d) P-polarised output power behaves for a single-pass (blue), forward double-pass (green), cavity (red) and reverse double-pass (cyan) discharge inside 140 cm of fibre B.

As before with the gain saturated by the double-pass, the transition from forward double-pass to cavity reduces the unpolarised and S-polarised output powers whilst increasing in the P-polarisation. There is however a difference with the single-pass result to before, with the transition to reverse double-pass having little to no effect. This is due to fortuitous positioning of a new sapphire window in the cathode gas cell resulting in a small back reflection of light into the fibre, turning what should be a single-pass measurement into an effective reverse double-pass measurement.

Figure 7.8e are measurements taken from both ends of the discharge with all mirrors blocked which show a significant asymmetry in the output powers due to this affect. Note that the cathode output is reflected from the 50/50 beamsplitter, while the anode output was coupled into the patch cord via two silver planar mirrors.

7.2.2   Modal Analysis

A cavity length of 160 cm (140 cm of fibre and two 10 cm radius of curvature mirrors) and an effective fibre modal index of very close to (but slightly below) 1 [4] give an estimated cold cavity mode spacing of 94 ± 2 MHz. If the broadened gain bandwidth is greater than this, the observed pulses should exhibit a beating modulation.

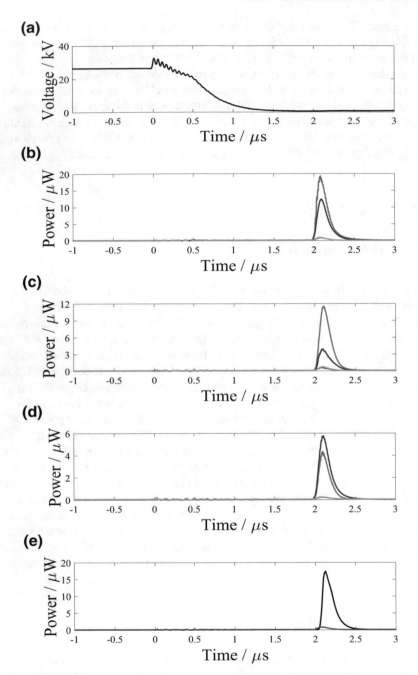

Fig. 7.8 Oscilloscope traces averaged over 128 pulses of 5:1 helium-xenon discharges at 12 mbar inside 140 cm of fibre B showing **a** the discharge voltage, **b–d** the unpolarised, S-polarised and P-polarised output powers respectively for the SP (blue line), FDP (green line), C (red line) and RDP (cyan line) cavity arrangements and **e** the SP powers from the anode end (black line) and cathode end (magenta line) of the fibre

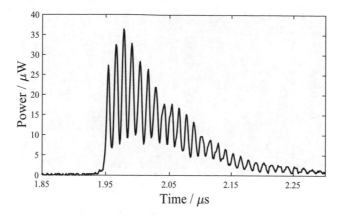

Fig. 7.9 Oscilloscope trace of a single output pulse from a 160 cm cavity

Equation 2.2 was used to calculate an inhomogeneous Doppler bandwidth of 93 ± 5 MHz, where for xenon $m = 2.18 \times 10^{-25}$ kg, $k_B = 1.38 \times 10^{-23}$ m^2 kg s^{-2} and $c = 3.00 \times 10^8$ m s^{-1} are all known constants, $f_0 = 8.57 \times 10^{13}$ Hz for the 3.5 μm line and the neutral xenon atom temperature in the afterglow is estimated to be 300 ± 30 K.

The homogeneous collisional broadening of a He-Xe mixture is much more difficult to calculate analytically. Instead an empirical estimate is made based on the work of Vetter and Marié [5], where the 3.5 μm line's collisional bandwidth for a 10 mTorr pressure of xenon in a mixture of helium was found to be

$$\Delta f_{pr} = 4.7 + (18.6 \pm 0.7)p, \tag{7.2}$$

where p is the pressure in Torr. For a pressure of 12 mbar (9 Torr) this gives a broadening of 172 ± 6 MHz. It is important to note however that this neglects any Xe-Xe collisions, and so is only a rough estimate.

Together the two broadening mechanisms result in a gain bandwidth of 265 ± 8 MHz, and so there should be multiple modes present in the laser cavity. The beating pattern of these is very clearly observed in Fig. 7.9, which is a single shot measurement of an output pulse from the aligned cavity.

The beating frequency can be retrieved by Fourier transforming the temporal measurement of the pulse. Averages were taken in an attempt to reduce the noise level, but as the phase of each pulse is both varying and unknown the frequency information is lost when averaging in the time domain. This is why the beating has been almost entirely smoothed out in Fig. 7.8. Instead, the Fourier transforms of 25 individual pulses were taken before their magnitudes where averaged to produce the plot in Fig. 7.10.

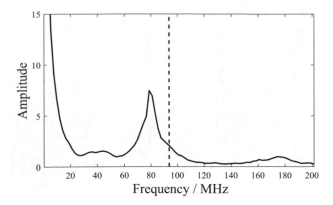

Fig. 7.10 Average of the Fourier spectra from 25 light pulses. The dashed line indicates the mode separation frequency for a 160 cm long cold cavity

Table 7.2 Breakdown of the loss of all components in the cavity

Component	Individual loss/dB	Cavity loss contribution/dB
Sapphire window	0.55	1.1
CaF$_2$ window	0.23	0.46
Silver mirror	0.12	0.24
Fibre transmission	0.16 dB m^{-1}	0.45
Fibre bend loss	0.2 dB turn^{-1}	0.4 (est.)
Output coupler	1.62	3.24
Total	–	5.89

There is a clear peak at 80 ± 2 MHz, a shift of almost 10 MHz from the cold cavity estimate (marked by the dashed black line). This frequency pulling means that an estimate for the saturated gain coefficient can be made by rearranging Eq. 2.8 to

$$\alpha_1 = \frac{\pi \Delta f}{l} \left[\frac{1}{\Delta v'} - \frac{1}{\Delta v} \right]. \tag{7.3}$$

Inserting the calculated values for Δf and Δv and the measured values for l and $\Delta v'$ gives $\alpha_1 = 1.1 \pm 0.2$ m^{-1}, and hence an amplification factor of $A = 22 \pm 10$ per round trip. In a lasing cavity A is equal to the loss, and so the estimated loss of the cavity is 14 ± 2.5 dB. Summing up the specified losses of all the components in the cavity yields an ideal cavity loss of 5.89 dB (see Table 7.2 for a detailed breakdown, note that in one round trip the light traverses each component twice). From this it can be inferred that there is an 8 ± 2.5 dB coupling loss from the mirrors, with each mirror coupling back $40 \pm 10\%$ of the light exiting the fibre.

To further test the frequency pulling, the anode mirror was defocused by around ~ 1 mm. This led to an increase in the cavity loss which would force the saturated

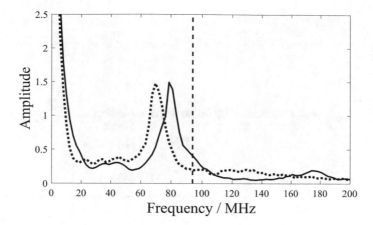

Fig. 7.11 Average of the Fourier spectra from 25 light pulses for a cavity with a focused (solid line) and a defocused (dotted line) anode mirror. The focused spectra amplitude has been reduced by a factor of 5 for an easier visual comparison. The dashed line indicates the mode separation frequency for a 160 cm long cold cavity

gain to increase, hence resulting in a stronger frequency pulling. Figure 7.11 shows that the mode beating frequency was reduced to 70 ± 2 MHz, a further shift of 10 MHz. As this was considerably more than the cold cavity frequency shift of around ~ 0.05 MHz caused by the small increase to the cavity length from the mirror movement, the shift must be a frequency pulling affect.

The same calculations can be done for this defocused case giving new values of $\alpha_1 = 2.2 \pm 0.3$ m^{-1} and $A = 400 \pm 300$, with a new total cavity loss of 26 ± 3 dB. This means that the loss at the anode mirror has increased from 4 ± 1 dB to 15 ± 3 dB; a coupling efficiency of $3 \pm 0.5\%$.

7.2.3 Evidence of Lasing on Alternative Lines

While the preliminary experiments showed strong evidence of gain on the 3.11 and 3.36 μm lines, no response from the mirrors had been seen in the CW measurements. However thanks to an easier and better alignment, when these wavelengths were measured using the monochromator for the same cavity as the previous section a response was now observable.

The 3.36 μm line exhibited the higher output power of the two, allowing for measurements in the unpolarised, S-polarised and P-polarised cases (Figs. 7.12a, b and c respectively). While the response is not as pronounced as for the main 3.51 μm line, the polarisation dependence again shows that oscillation is occuring inside the cavity at this wavelength. With the lower output power for the 3.11 μm line in Fig. 7.12d measurements in the P-polarisation were not possible, but the unpolarised trace does demonstrate some mirror response.

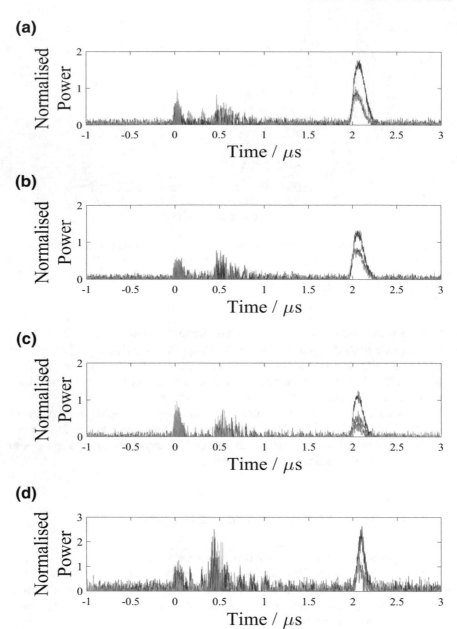

Fig. 7.12 Oscilloscope traces averaged over 64 pulses of 5 : 1 helium-xenon discharges at 12 mbar inside 140 cm of fibre B showing **a–c** the unpolarised, S-polarised and P-polarised output powers at 3.36 μm respectively normalised to the unpolarised single-pass peak and **d** unpolarised output power at 3.1 μm normalised to the single-pass peak for the single-pass (blue line), forward double-pass (green line), cavity (red line) and reverse double-pass (cyan line) cavity arrangements

The low output powers also made the observation of cavity modes difficult, and any measurements made gave unstable results.

7.3   Pressure Optimisation

Up until now, all the experiments described were performed in a 5:1 helium-xenon ratio at a pressure of 12 mbar. As this was based on previous research in continuous discharges with larger diameter discharge tubes, some investigation into whether the performance of the hollow core fibre based laser could be improved by changing the gas pressure or mixture was required.

Four different He-Xe ratios were tested in 100 cm of fibre B; 5:1, 7:1, 10:1 and 20:1. At each ratio measurements were made at increasing pressures starting at the lowest pressure at which breakdown could be achieved, before a final control measurement at 12 mbar was made to ensure that no external degradation had occurred. At each pressure the output power for the single-pass, forward double-pass, cavity and reverse double-pass arrangements were all recorded as well as the repetition rate of the pulsing. All measurements were made with a measured drawn current from the power supply of 0.25 mA. The results of these experiments can be seen in Fig. 7.13.

Figures 7.13a, b plot the peak power against pressure for a single-pass and cavity respectively. They show a clear inverse relation between the gas pressure and the output power, with smaller partial pressures of xenon also yielding higher powers. The correlation is stronger in the single-pass case, where there is no external influence from the natural drift of the cavity mirrors over the time of the measurements.

Figures 7.13c, d feature the complete cavity output power traces for all the pressures at a 7:1 ratio, and for ratios of 5:1, 7:1 and 10:1 at 12 mbar respectively. As well as a decrease in peak power at higher pressures, they also show that the pulses are delayed further with both higher total and partial xenon pressures. It is not clear from this data whether these two factors are related. It is possible that at lower pressures the electron temperature settles to the opportune value faster, and so recombination to the upper lasing levels is more efficient and the upper state population is higher when emission begins to occur. Further experimentation into the temperature and state populations would be needed to know this for sure. Whether related or not, the results appear contradictory to the general idea that high pressures are required to produce high powers in gas lasers. It is possible that two different regimes exist for low and high pressure.

Figures 7.13e, f plot the pulse frequency and FWHM against pressure respectively. The pulse frequency exhibits no dependence on partial pressure, and only a very small inverse dependence of total pressure. Conversely, the pulse duration exhibits a small positive dependence on pressure. However neither of these relations are particularly strong, and so the variation of the laser's average power will closely follow that of the peak power.

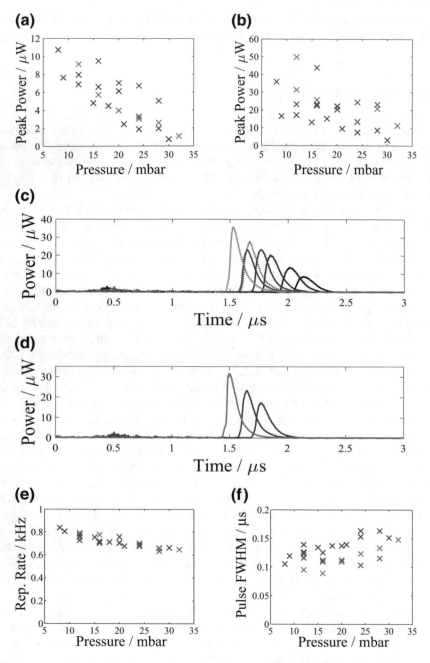

Fig. 7.13 Pressure optimisation of the He-Xe laser in 100 cm of fibre B at gas ratios of 5:1 (red), 7:1 (blue), 10:1 (green) and 20:1 (magenta). **a** single-pass peak power **b** cavity peak power **c** cavity output power traces for a 7:1 mixture at pressures of 8, 12, 16, 20, 24 and 28 mbar (light to dark, note the dotted 12 mbar measurement was a control taken last) (c) cavity output power traces at 12 mbar for different mixtures **e** pulse repetition rate **f** light pulse duration (FWHM)

7.4 Summary

In this chapter it has been demonstrated that the discharges were pulsing in semi-stable manner akin to that of a Pearson-Anson oscillator thanks to a weak stray capacitance around the high voltage anode cell. These current pulses were found to be accompanied by a strong afterglow light pulse, mostly from the 3.51 μm xenon line but also including light from the 3.11 and 3.36 μm xenon lines.

Aligning the cavity mirrors elicited a response from the light pulse similar to the CW measurement, with lasing again confirmed by a polarised cavity output. This was observed on both the 3.36 and 3.51 μm xenon lines, with a small response also detected on the 3.11 μm line. Lasing was also confirmed by the existence of mode beating. The high gain in the lasing medium induced frequency pulling effects, and so estimations for the saturated gain and cavity loss of 1.1 ± 0.2 m^{-1} and 14 ± 2.5 dB respectively could be made.

Finally it was found that decreasing the total pressure and partial xenon pressure increased the power of the afterglow pulse. As previous systems have shown that high pressures produce high powers [6–9], this seemed a strange result. However these high power systems are based on the 1.73 and 2.02 μm xenon lines meaning the mid-IR lines measured might not perform the same.

References

1. P. Persephonis, V. Giannetas, C. Georgiades, J. Parthenios, A. Ioannou, The inductance and resistance of the laser discharge in a pulsed gas laser. IEEE J. Quantum Electr. **31**(3), 573 (1995)
2. D. R. Lide, *CRC Handbook of Chemistry and Physics*, CRC Press (2008)
3. E. B. Saloman, Energy levels and observed spectral lines of Xenon, Xe I through Xe LIV (2004)
4. E.A.J. Marcatili, R.A. Schmeltzer, Hollow metallic and dielectric waveguides for long distance optical transmission and lasers. Bell Syst. Techn. J. **43**(4), 1783 (1964)
5. E. Marié, R. Vetter, Phase-changing broadening of the laser line of Xe I at $\lambda=3.51\mu m$. J. Phys. B: Atomic Mol. Phys. **11**(16), 2845 (1978)
6. S.E. Schwarz, T.A. DeTemple, R. Targ, High-pressure pulsed xenon laser. Appl. Phys. Lett. **17**(7), 305 (1970)
7. S.A. Lawton, J.B. Richards, L.A. Newman, L. Specht, T.A. DeTemple, The high pressure neutral infrared xenon laser. J. Appl. Phys. **50**(6), 3888 (1979)
8. L.N. Litzenberger, D.W. Trainor, M.W. McGeoch, A 650 J e-beam pumped atomic xenon laser. IEEE J. Quantum Electr. **26**(9), 1668 (1990)
9. J.P. Apruzese, J.L. Giuliani, M.F. Wolford, J.D. Sethian, G.M. Petrov, D.D. Hinshelwood, M.C. Myers, A. Dasgupta, F. Hegeler, T. Petrova, Optimizing the ArXe infrared laser on the Naval Research Laboratory's Electra generator. J. Appl. Phys. **104**(1), 13101 (2008)

Chapter 8
Experiments with New Gas Mixtures

Having demonstrated a fully functional afterglow helium-xenon laser, attention was turned to the use of helium-neon. These helium-neon experiments were unsuccessful, however they did demonstrate that using neon instead of helium had the potential to improve the xenon laser.

The final part of this chapter sees these gases replaced by argon, for the investigation of the emission from both the neutral and singly ionised argon atom. This chapter presents experiments at an early stage of completion due to time and equipment restrictions which require further investigation.

8.1 Neon Experiments

8.1.1 Helium-Neon

The starting pressures for helium-neon discharges in hollow fibres were again based on previous results in the literature. Willett quotes an optimum mixture of $5:1$ helium to neon with a pD product between 4.8 and 5.3 mbar mm for oscillation on the 632.8 nm line [1]. However Willett also references Fields [2], who showed that for decreasing discharge tube diameters the optimum He-Ne ratio shifts to small partial pressures of neon. This is backed up by Smith's He-Ne waveguide laser, where a $10:1$ ^{3}He to ^{20}Ne ratio at a pD of 4 mbar mm yielded the highest gain [3]. The slightly lower than expected pD may result from the isotopes of helium and neon used, or an indication that the optimum pD drops for very small diameters, as for xenon in the previous chapter.

Unfortunately the much higher ionization energy of neon compared to xenon places much harsher restrictions on the pressures at which a discharge is possible, and in a 55 cm fibre the lowest achievable ratio and pressure was $5:1$ at 42 mbar; a pD of 4.8 mbar mm.

Before testing helium-neon, the fibre was filled with helium-xenon for a small number of discharges so that the optics could be pre-aligned. After 24 h of being

© Springer International Publishing AG, part of Springer Nature 2018
A. Love, *Hollow Core Optical Fibre Based Gas Discharge Laser Systems*,
Springer Theses, https://doi.org/10.1007/978-3-319-93970-4_8

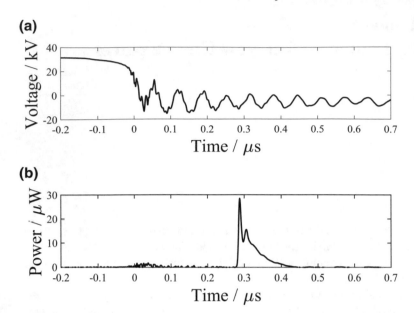

Fig. 8.1 Oscilloscope traces averaged over 128 pulses of 5:1 helium-neon discharges with trace amounts of xenon at 42 mbar inside 55 cm of fibre B showing **a** the discharge voltage and **b** the output power

under vacuum, the fibre was then filled with the helium-neon mixture and discharged. The resulting traces can be seen in Fig. 8.1. On first glance it appears that the helium-neon system is producing a large amount of light, but based of the colour of the discharge (seen in Fig. 8.2a) it seems that despite a long evacuation there are still trace amounts of xenon within the fibre. The spectral location of these light pulses was later verified to indeed be at 3.5 μm, not 3.39 μm where the neon line exists.

It should be noted that the discharge voltage trace is significantly different to the previous helium-xenon discharges, which has come from either the new gas mixture or the shorter length. One minor cause for concern is the negative values of the measured voltage after the discharge has struck, however this is more likely to be another artefact of the far insufficient bandwidth of the voltage probe.

Figure 8.2b is a photograph of a similar helium-neon discharge after the fibre has been evacuated and flushed with neon a few times so that the trace xenon has been flushed out. In this case the fibre no longer has any blue xenon glow, but is entirely glowing with the red-orange colour of neon. Unfortunately the xenon colouring was not the only thing lost, as there was no longer any signal detected either.

The lack of signal is likely to be due to the high pressures used in the discharges. While a direct comparison to the helium-xenon case cannot be made due to the differing excitation methods, the output power from a 5:1 mixture of helium-xenon was reduced to zero before reaching 35 mbar of gas, a lower pressure than that of the 42 mbar of helium-neon.

(a)

(b)

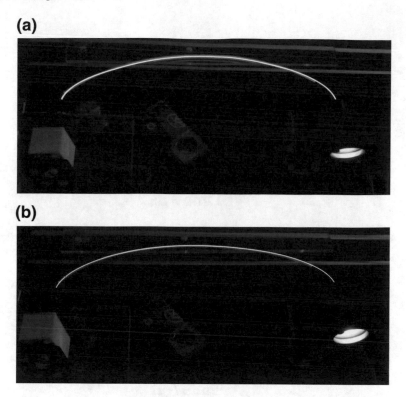

Fig. 8.2 Photographs of a 55 cm He-Ne discharge (**a**) with trace amounts of Xe and (**b**) after a number of flushes with Ne to remove any remaining Xe

Unfortunately as the ability to achieve a discharge placed a minimum limit on the available pressures, an operable helium-neon laser would require an overhaul of the system. The were a number of options for this; redesigning the electrode to reduce the fall potentials, introducing some high voltage trigger to initiate the discharge (like a spark gap) or a total redesign to establish a true CW system where working pressures may be higher (see Chap. 9 for suggestions). However these options were all outside of the remaining timescale of the project.

8.1.2 *Helium-Neon-Xenon*

While the xenon contamination in the previous section was undesirable at the time, the high output power from such a short discharge showed that the helium-xenon laser could be improved upon. It had been previously shown in Chap. 7 that a lower partial pressure of xenon produced a higher output power, but stable discharges became harder to achieve. The introduction of the neon however seems to have stabilised

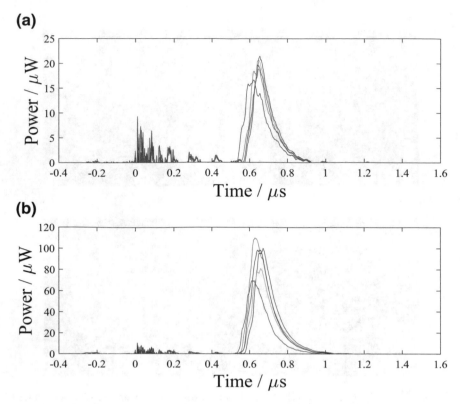

Fig. 8.3 Oscilloscope traces of the output power from He-Ne-Xe discharges inside 80 cm of fibre A showing the (**a**) single-pass and (**b**) cavity arrangements with 0.1 mbar of Xe and 12 mbar of He-Ne at ratios of 4:2 (blue), 3:3 (green), 2:4 (red), 1:5 (cyan) and 0:6 (magenta). All traces are taken with an averaging of 128

the discharges and subsequently reduced the previous restrictions on the minimum partial pressure.

To investigate the effect of adding neon to a helium-xenon mixture, the output power from an 80 cm long piece of fibre A was monitored for varying mixtures of helium and neon with a xenon partial pressure of 0.1 mbar and a total pressure of 12 mbar. Figure 8.3a and b are the output power traces for the single-pass and cavity respectively for He-Ne ratios of 4:2, 3:3, 2:4, 1:5 and 0:6. A 5:1 ratio discharge was attempted, but the stability was insufficient to record the signal.

The traces show little sign of a correlation between the helium-neon ratio and the output power, suggesting that the neon is just taking on the same role as helium; introducing more electrons, producing more xenon ions through Penning collisions and quenching the lower lasing levels. Note that as with helium, there are no neon energy levels coincident with the xenon upper lasing levels. The discharges do however become more stable as the partial pressure of neon increases. The reason for this is quite straightforward; the ionisation energy of neon is lower than that of helium.

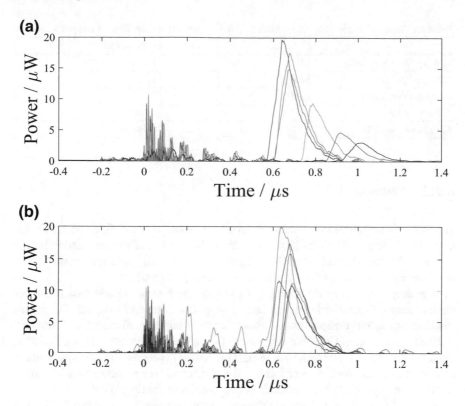

Fig. 8.4 Oscilloscope traces of the single-pass output power from Ne-Xe discharges inside 80 cm of fibre A at pressures of 12 (blue), 18 (green), 24 (cyan), 30 (magenta), 36 (black) and 12 mbar (orange) for (**a**) a fixed 120 : 1 Ne-Xe ratio and (**b**) a fixed 0.1 mbar of Xe. All traces are taken with an averaging of 128

With easier discharges and no real affect on the output power, all subsequent experiments were conducted with just neon and xenon. However before proceeding with the characterisation of the laser medium cavity a short investigation was made into the impact of the partial xenon pressure against the total pressure on the output.

The traces in Fig. 8.4a are of the single-pass output with various total pressures at a set Ne-Xe ratio of 120 : 1 while those in Fig. 8.4b have a fixed xenon pressure of 0.1 mbar. Perhaps unsurprisingly when the ratio is fixed the output power degrades with increasing pressure, but when doing so with a fixed xenon pressure the degradation is significantly less. This implies the partial pressure of xenon is the more important factor and that Xe-Xe and Xe-wall collisions have a much bigger impact on the output power.

Table 8.1 Loss calculation for the 85 cm cavity with the anode mirror focused and unfocused

	Focused	Defocused
Beating frequency/MHz	150 ± 7	127 ± 10
Saturated gain/m^{-1}	0.8 ± 0.3	1.7 ± 0.5
Amplification factor	3 ± 1	9 ± 6
Calculated loss/dB	4 ± 2	10 ± 3
Component loss/dB	5.49	5.49

8.1.3   Neon-Xenon

A new cavity was constructed with a fresh 65 cm piece of fibre A, filled with 10 mbar of neon and xenon at a ratio of 200 : 1. The cavity was characterised in a similar fashion to the helium-xenon case; first an investigation into the cavity polarisation was made, followed by an analysis of the cavity mode beating frequencies.

The single-pass, forward double-pass, cavity and reverse double-pass traces for each of the S-, P-polarised and unpolarised outputs are shown in Fig. 8.5. Other than the single-pass and reverse double-pass trace now having a discernible difference, the behaviour is essentially the same as in Chap. 7 which means that oscillation is occurring. The difference in the single-pass behaviour is because the sapphire window at the cathode had been replaced and so its alignment was no longer coincident to the fibre, removing any back reflections that were contributing to the cavity.

Having replaced the helium with neon, the pressure broadening bandwidth becomes

$$\Delta f_{pr} = 4.7 + (8.3 \pm 0.5)p, \tag{8.1}$$

and so the neon-xenon bandwidth is 160 ± 6 MHz [4]. With a shorter cavity length of 85 cm, there is an expected cold cavity mode separation of 176 ± 5 MHz and so unless there is a considerable amount of frequency pulling there will only be at most two longitudinal modes. This is reflected in the much weaker mode beating pattern observed in Fig. 8.6a.

As in Chap. 7 single traces of the light pulses were Fourier transformed before their magnitudes were averaged for both the focused and defocused anode mirror cases to obtain the plots in Fig. 8.5b. These could then be used to calculate the saturated gain and cavity loss, outlined in Table 8.1.

While the component loss does fit inside the uncertainty for the calculated loss, something does seem to be amiss as the component loss is the absolute minimum possible when there is 100% coupling by the mirrors. The most likely cause of the discrepancy is the much lower homogeneous broadening. Either there is some extra broadening from elsewhere that is not taken into account, or the more inhomogeneous broadening reduces the effect of the frequency pulling as the earlier calculations are based on a homogeneous gain bandwidth.

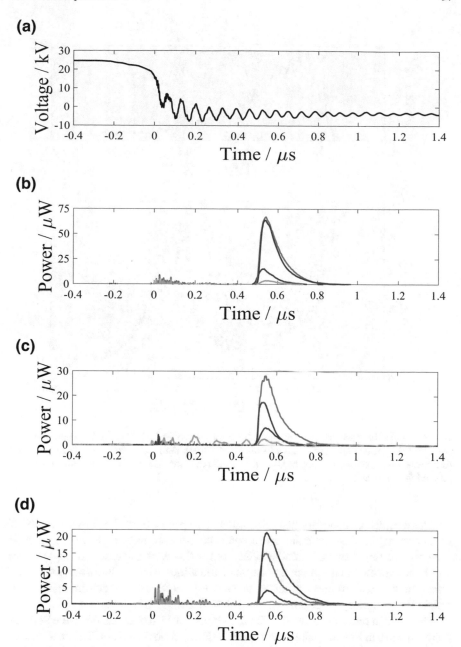

Fig. 8.5 Oscilloscope traces averaged over 128 pulses of 200:1 neon-xenon discharges 10 mbar inside 65 cm of fibre A showing (**a**) the discharge voltage, (**b**)–(**d**) the unpolarised, S-polarised and P-polarised output powers respectively for the SP (blue line), FDP (green line), C (red line) and RDP (cyan line) cavity arrangements. The noisy RDP trace in (**c**) is a result of a slight discrepancy between the 'light' and 'dark' traces, which is magnified by the subtraction of the dark trace

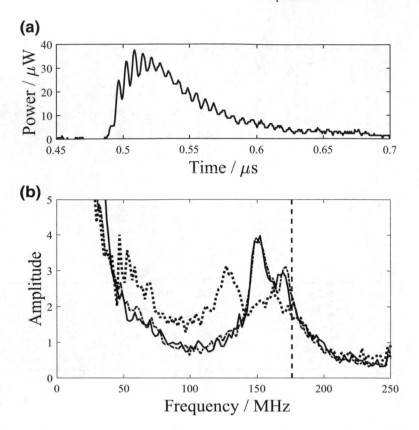

Fig. 8.6 **a** Oscilloscope trace of a single pulse from the cavity. **b** Average of the Fourier spectra from 20 light pulses for a cavity with a focused (solid line), defocused (dotted line) and refocused (dot-dashed line) anode mirror. The dashed line indicates the mode separation frequency for an 85 cm long cold cavity

There is also a secondary peak in the Fourier spectrum for the focused cavity of unknown origin. One potential cause could be some reflections from the cathode window creating a second, smaller cavity, but if this were the case a stronger effect would be expected in the previous case when the window had a more obvious impact. One potential way to clear up these issues would be to use a spectrum analyser to obtain the frequency spectrum over a much wider range of pulses.

The 3.5 μm emission as well as small 3.36 and 3.11 μm emissions were verified by the spectra in Fig. 8.7, taken using the Bentham monochromator. The spectra also identified a strong emission at 2.02 μm that had not previously been investigated.

The behaviour of the 2.02 μm line was monitored using a second InGaAs detector (ThorLabs PDA10D, 1.2 to 2.6 μm wavelength range, 15 MHz bandwidth, 1.3 A W^{-1} responsivity at 2.3 μm, 5 kV A^{-1} gain). The traces in Fig. 8.8 curiously show a strong pulse at the onset of the discharge, terminating before the main current

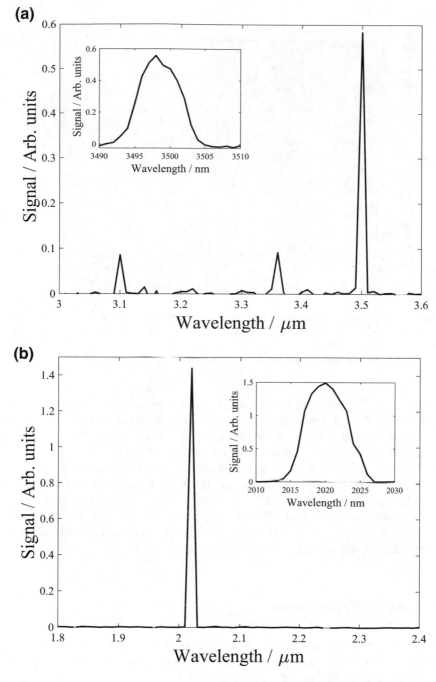

Fig. 8.7 Spectra of the cavity output signal from 200 : 1 neon-xenon discharges at 10 mbar inside 65 cm of fibre A (**a**) in the mid-IR range using an InSb detector and (**b**) using a PbSe detector with a resolutions of 10 nm. Insets: the (**a**) 3.51 μm and (**b**) 2.02 μm lines zoomed in with a higher number of data points. The mid-IR peak is shifted from 3.51 μm due to a slight miscalibration of the spectrometer

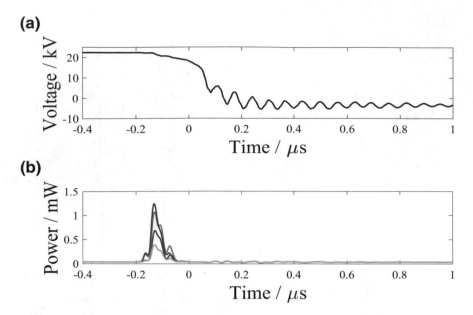

Fig. 8.8 Oscilloscope traces averaged over 128 pulses of 200:1 neon-xenon discharges at 10 mbar inside 65 cm of fibre A showing (**a**) the discharge voltage and (**b**) the unpolarised output power from an amplified InGaAs detector sensitive to 2 μm for the SP (blue line), FDP (green line), C (red line) and RDP (cyan line) cavity arrangements

pulse begins. However, once again without more sophisticated equipment no real conclusions about this emission can be made. The low bandwidth of the detector also means that these pulses cannot be probed for mode beating, but the response from each of the mirrors indicates that the line is in fact lasing.

The neon-xenon gas mixtures are not without their drawbacks however. While the 12 mbar 5:1 He-Xe mixtures did not display any cataphoretic gas separation over the discharge lifetimes, with a much larger ratio of neon to xenon some separation can be observed. These effects are still not particularly significant however, and it is only in discharges with timescales of up ~10 min that the effect can be observed both in the colouration and stability of the discharge.

8.2 Argon Experiments

To investigate the potential of an argon gas discharge as a lasing medium the helium bottle connected to the vacuum system was replaced by argon (N5.5 purity). A picture of the resulting argon glow can be seen in Fig. 8.9.

The four IR lines of neutral argon around 2 μm were investigated with the amplified InGaAs detector used in the preceding section at a variety of argon pressures

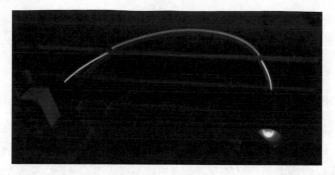

Fig. 8.9 Photograph of a 50 cm Argon discharge

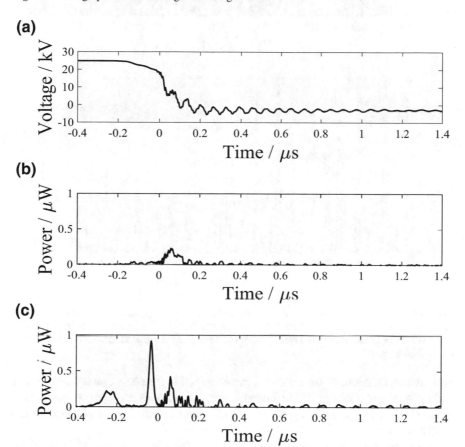

Fig. 8.10 Oscilloscope traces averaged over 128 pulses of argon discharges inside 50 cm of fibre A showing (**a**) the discharge voltage and the output powers from an amplified InGaAs detector sensitive to 2 μm at (**b**) 10 mbar and (**c**) 5 mbar. No spectral selection was used due to the very low powers detected

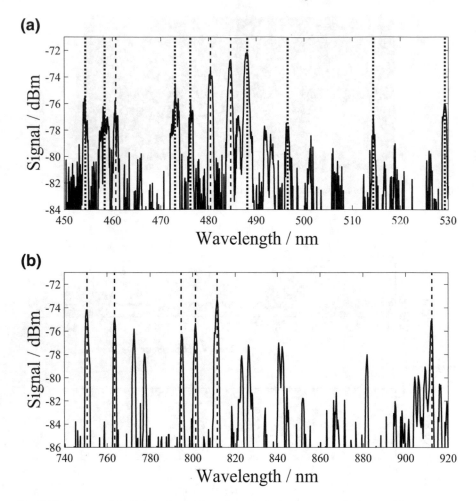

Fig. 8.11 Spectra of the light collected from the end of a 5 mbar argon discharge with a resolution of 1 nm. Emission lines that are known to lase are marked with dotted lines, notable non-lasing lines with dashed lines. Additional emission lines of neutral argon are present but unmarked as they are of little interest

and mixes with neon. While no signal was observed in the Ne-Ar mixes, Figs. 8.10b and c, where pressures of 10 and 5 mbar of pure argon were used respectively, show small signs of a signal. However these signals demonstrated no response from the cavity mirrors.

With a current pulse in excess of 1000 A cm^{-2} there is a real possibility of lasing on the argon ion lines. As with the xenon discharges in Chap. 7 the discharge output was fed into an OSA so that the emissions lines present could be identified, and their intensities compared (see Fig. 8.11). A selection of notable lines observed are listed in Table 8.2. As with xenon the emission intensities between neutral and singly ionised argon cannot be directly compared, but with approximately equal measured

Table 8.2 A selection of strongly observed lines in an Ar discharge as measured by a Yokogawa AQ-6315A optical spectrum analyser with a resolution of 1 nm. The accepted values, species and strengths are taken from [5] except for entries marked with * which come from [6]. Lines that are known to lase in other systems have been labelled accordingly

Observed line/nm	Accepted wavelength/nm	Specie	Relative strength	Lases
454	454.505	Ar II	400	✓
458	457.935	Ar II	400	✓
461	460.957	Ar II	550	✗
473	472.686	Ar II	550	✓
476	476.487	Ar II	800	✓
480	480.602	Ar II	550	✗
485	484.781	Ar II	150	✗
488	487.986	Ar II	800	✓
497	496.508	Ar II	200	✓
514	514.531	Ar II	70	✓
529	528.689*	Ar II	N/A	✓
750	750.387	Ar I	20000	✗
764	763.510	Ar I	25000	✗
795	794.818	Ar I	20000	✗
801	801.479	Ar I	25000	✗
812	811.531	Ar I	35000	✗
912	912.297	Ar I	35000	✗

strengths when the neutral lines are quoted as being significantly stronger it is safe to assume that there is a high population of ionised atoms.

Unfortunately though, there was no detectable response from the cavity mirrors and so no evidence of gain on any of the lines. This was not in fact surprising, as the loss of the fibres (particularly the bend loss) in this wavelength range is high. Also as with the helium-neon the pressure may be incorrect, or the pulse current could be wrong.

These experiments however have demonstrated that a fibre on this scale can withstand the high current densities required. A fibre with better guidance in the visible or a more controlled pulsing regime (or CW) could potentially demonstrate lasing on these lines.

8.3 Summary

In this chapter different gas mixtures have been shown to support discharges including helium-neon, neon-xenon and argon. Any signal from the helium-neon discharges in the mid-IR were shown to actually be from trace amounts of xenon leftover in the fibre.

It was demonstrated that discharges could be achieved at lower partial pressures of xenon using neon as the buffer gas instead of helium. This resulted in output powers comparable to those of the helium-xenon discharges in the previous chapter that were over twice the length. The subsequent neon-xenon laser exhibited the same polarisation and mode beating properties as the helium-xenon laser.

In the argon discharges there was little light detected in the neutral lines with a fast detector, but moderate levels of light detected on the ion laser line in a slow spectral measurement. No response was reported from the cavity mirrors in either case.

References

1. C.S. Willett, *Introduction to Gas Lasers: Population Inversion Mechanisms* (Pergamon Press, 1974)
2. R.L. Field Jr., Operating parameters of dc-excited He-Ne gas lasers. Rev. Sci. Instrum. **38**, 1720 (1967)
3. P.W. Smith, A waveguide gas laser. Appl. Phys. Lett. **19**, 132 (1971)
4. E. Marié, R. Vetter, Phase-changing broadening of the laser line of Xe I at $\lambda = 3.51$ μm. J. Phys. B At. Mol. Phys. **11**(16), 2845 (1978)
5. D.R. Lide, *CRC Handbook of Chemistry and Physics* (CRC Press, 2008)
6. E.B. Saloman, Energy levels and observed spectral lines of ionized argon, Ar II through Ar XVIII. J. Phys. Chem. Ref. Data **39**(3) (2010)

Chapter 9
Conclusions and Future Prospects

The research presented within this thesis is experimental evidence of the first functional hollow core fibre based electrically pumped gas laser. Taking the form of an afterglow xenon laser, this experiment is primarily a proof of concept that opens the door to a new generation of compact, flexible gas discharge laser systems.

In Chap. 6 it was demonstrated that electrical discharge could be achieved in a 5:1 mixture of helium-xenon at a pressure of 12 mbar contained within 1.5 m of a 120 μm ID hollow core fibre, far surpassing the previous record of 13.7 cm in a 150 μm ID capillary [1]. A cavity was constructed around these fibres using two silver spherical mirrors. A signal increase greater than a factor of two on the 3.51 μm xenon line from a forward double-pass set up confirmed the existence of gain, while a similar reduction in signal indicating gain saturation was detected from a reverse double-pass measurement. With an output coupler that exhibits 60% reflection in the S-polarisation and 30% reflection in the P-polarisation, the laser cavity became primarily P-polarised and a much higher output signal in the P-polarisation showed that oscillation was occurring.

In Chap. 7 it was established that the discharges were not continuous as previously indicated, but in fact were pulsing in a semi-stable manner thanks to a weak stray capacitance around the high voltage anode gas cell. This capacitance was calculated to be ~10 pF, giving an estimate for the current flowing through the discharge with each pulse to be ~1 A. The pulse repetition rate was found to be around 1 kHz, although this did vary with the drawn current. The light from mid-IR xenon lines was found to be mostly located ~1 μs after the current pulse had ceased and hence in the afterglow. The much better signal-to-noise ratio achievable from measuring the afterglow pulse of ≈200 ns in duration enabled better mirror alignment. Response from the mirrors was observed on the 3.11, 3.36 and 3.51 μm xenon lines, and the subsequent laser cavity output exhibited the polarisation dependence on both the 3.36 and 3.51 μm lines. The cavity output pulse also displayed some modulation from cavity mode beating effects. In a 1.6 m long cavity the frequency of this beating was found to be 80 MHz, some 10 MHz lower than the cold cavity frequency of

© Springer International Publishing AG, part of Springer Nature 2018
A. Love, *Hollow Core Optical Fibre Based Gas Discharge Laser Systems*,
Springer Theses, https://doi.org/10.1007/978-3-319-93970-4_9

93 MHz thanks to significant frequency pulling effects. From this a saturated gain of $1.1\,m^{-1}$ was calculated, giving a cavity loss of 13.5 dB. The frequency could be pulled even lower by defocusing the mirror and increasing cavity loss. Finally it was found that a higher output power could be achieved by decreasing the total working pressure, and the xenon to helium ratio. Unfortunately the lower pressures did make breakdown harder to achieve, effectively placing a minimum value on the usable pressure.

In Chap. 8 a number of different gas mixtures were shown to support discharges, including helium-neon, neon-xenon and argon. In the helium-neon mixes no signal at the 3.39 μm neon line was detected, however trace amounts of leftover xenon gave a strong emission at 3.51 μm comparable to that of the previous discharges but using half the length of fibre. It was found that by using neon instead of helium lower partial pressures of xenon could be used in discharges and higher output powers could be achieved. Using a 10 mbar mixture of neon-xenon at a ratio of 200:1 a laser cavity produced a comparable output power to the 1.6 m cavity in the previous chapter in a cavity only 85 cm in length. The neon-xenon laser demonstrated the same polarisation and mode beating properties as the helium-xenon laser. The argon discharges produced limited light from the neutral laser lines, and moderate signal from the ion laser lines. While no mirror response was detected on these ion lines, a fibre with better visible guidance could function better.

While this research has well and truly laid the groundwork for fibre gas discharge lasers, there is still much to be done. One of the biggest issues facing the system as it stands is the unstable and ill-defined nature of the pulsing. Without a suitable high speed voltage probe a lot of assumptions have been made about the current pulses, which could be avoided if there was more control over the discharge. One possible solution would be to redesign the anode gas cell so as to remove as many sources of the stray capacitance as possible. The surrounding metal fittings could be replaced without too much difficulty, but the stainless steel Swagelok valve directly connected to cell would be an engineering challenge. For the same reasons as the cell itself, a plastic valve is not suitable and so it would need to be made of ceramic or glass. On a somewhat related note, a 3D printed fused silica gas cell is an intriguing prospect [2].

A more plausible solution would be to combine the current transverse DC excitation with an RF or microwave excitation system. This is the method that was adopted by Smith in the early waveguide gas lasers to eliminate the 'relaxation-oscillation discharge instability in narrow-bore DC discharges' [3]. This would require little modification of the current system and so would be easy to implement.

Another way to develop these new gas lasers would be to investigate ways of improving their output power. One possibility is to increase the active volume by increasing the length of the fibre, while another would be to increase the active pressure and include argon. Both methods however would push the voltages required to achieve breakdown to potentially unreachable values. Previous systems [4–7] utilised transverse excitation methods, either with microwave discharges, DC discharges or an electron gun, which seem a more plausible route. The latter two of these would be very difficult to implement into the fibre design, but microwave excitation is a possibility.

References

1. X. Shi, X.B. Wang, W. Jin, M.S. Demokan, Investigation of glow discharge of gas in hollow-core fibers. Appl. Phys. B Lasers Optics **91**(2), 377 (2008)
2. F. Kotz, K. Arnold, W. Bauer, D. Schild, N. Keller, K. Sachsenheimer, T.M. Nargang, C. Richter, D. Helmer, B.E. Rapp, Three-dimensional printing of transparent fused silica glass. Nature **544**(7650), 337 (2017)
3. P.W. Smith, A waveguide gas laser. Appl. Phys. Lett. **19**, 132 (1971)
4. C.L. Gordon, C.P. Christensen, B. Feldman, Microwave-discharge excitation of an ArXe laser. Opt. Lett. **13**(2), 114 (1988)
5. S.A. Lawton, J.B. Richards, L.A. Newman, L. Specht, T.A. DeTemple, The high pressure neutral infrared xenon laser. J. Appl. Phys. **50**(6), 3888 (1979)
6. L.N. Litzenberger, D.W. Trainor, M.W. McGeoch, A 650 J e-beam pumped atomic xenon laser. IEEE J. Quantum Electron. **26**(9), 1668 (1990)
7. J.P. Apruzese, J.L. Giuliani, M.F. Wolford, J.D. Sethian, G.M. Petrov, D.D. Hinshelwood, M.C. Myers, A. Dasgupta, F. Hegeler, T. Petrova, Optimizing the ArXe infrared laser on the Naval Research Laboratory's Electra generator. J. Appl. Phys. **104**(1), 13101 (2008)